Riemannian Geometry
in an Orthogonal Frame

I0038349

Riemannian Geometry in an Orthogonal Frame

From lectures delivered by Élie Cartan
at the Sorbonne in 1926–27

Translated from Russian by

Vladislav V. Goldberg

New Jersey Institute of Technology
Newark, New Jersey, U.S.A.

Foreword by **S. S. Chern**

World Scientific
New Jersey • London • Singapore • Hong Kong

Published by

World Scientific Publishing Co. Pte. Ltd.

5 Toh Tuck Link, Singapore 596224

USA office: 27 Warren Street, Suite 401-402, Hackensack, NJ 07601

UK office: 57 Shelton Street, Covent Garden, London WC2H 9HE

British Library Cataloguing-in-Publication Data
A catalogue record for this book is available from the British Library.

RIEMANNIAN GEOMETRY IN AN ORTHOGONAL FRAME

Copyright © 2001 by World Scientific Publishing Co. Pte. Ltd.
All rights reserved. This book, or parts thereof, may not be reproduced in any form or by any means, electronic or mechanical, including photocopying, recording or any information storage and retrieval system now known or to be invented, without written permission from the publisher.

For photocopying of material in this volume, please pay a copying fee through the Copyright Clearance Center, Inc., 222 Rosewood Drive, Danvers, MA 01923, USA. In this case permission to photocopy is not required from the publisher.

ISBN-13 978-981-02-4746-1
ISBN-10 981-02-4746-X
ISBN-13 978-981-02-4747-8 (pbk)
ISBN-10 981-238-062-0 (pbk)

Foreword

Riemann's geometry was a generalization of Gauss's theory of surfaces. He introduced the curvature tensor, the sectional curvature, and derived the conformal form of the metric of constant curvature. The latter was the only formula in his paper.

It was recognized that the fundamental problem is the form problem, i.e., the condition for two Riemannian metrics to differ by a coordinate transformation. This was solved independently by Christoffel and Lipschitz in 1870, by different methods. Of these Christoffel's method was a forerunner of Ricci's tensor analysis.

The existence of an absolute differentiation in the tensor bundle in Riemannian geometry, as defined by Levi–Civita, is a beautiful and fundamental fact. However, a more basic analytic tool on manifolds is the exterior calculus introduced by Cartan in 1922, which exists as a result of the differentiable structure.

When we do analytical Euclidean geometry, we would prefer an orthogonal coordinate system, instead of a general Cartesian system. Cartan carried this out for Riemannian geometry. In this sense the book does not need any further recommendation.

S. S. Chern

Translator's Introduction

Élie Cartan's book *Leçons sur la géométrie des espaces de Riemann*[1], published in 1928, is one of his best introductory books on his methods. This book was based on his lectures of the academic year 1925–1926. In 1936 this book was translated into Russian.[2] A modernized and much enlarged second, French, edition appeared in 1946.[3] In this edition three new chapters (on symmetric spaces, Lie groups of motions, and isometric Riemannian manifolds) and two new notes were added. The second edition was translated into English[4] and augmented by notes and appendices by R. Hermann relating the work to the modern point of view.

My Ph. D. advisor, S. P. Finikov took notes at another series of Cartan's lectures on geometry of Riemannian manifolds given at the Sorbonne in 1926–1927. These lectures were different from the lectures of 1925–1926. In this course Cartan introduced exterior differential forms at the outset and used orthogonal frames throughout in his investigation of Riemannian manifolds. He solved a series of problems in Euclidean and non-Euclidean spaces and a series of variational problems concerning geodesics.

Initially, Finikov translated into Russian, for his students, only the part of his notes dealing with the method of exterior differential forms. Later he translated all of his notes and published them as *Rimanova geometriya v ortogonal'nom repere*.[5] Since only the first edition of Cartan's book was available in Russian, Finikov included translations of most (but not all) of the additional material from the second, French, edition. In 1964, this book was translated

---

[1]Cartan, É. *Leçons sur la géométrie des espaces de Riemann*, Gauthier-Villars, Paris, 1928, vi+273 pp. (JFM 54.0755.01)

[2]Cartan, É, *Geometriya Rimanovyx prostranstv*, translated by G. N. Berman, edited by A. M. Lopshits, ONTI, 1936, 244 pp.

[3]Cartan, É., *Leçons sur la géométrie des espaces de Riemann*, 2nd edition, Gauthier-Villars, Paris, 1946, viii+378 pp. (MR **8**, 602; Zbl. **60**, 381); 2nd printing, 1951 (MR **13**, 491; Zbl. **44**, 184); 3rd printing, 1988.

[4]Cartan, É., *Geometry of Riemannian spaces*, translated from the French and with a preface by J. Glazebrook; with a preface, notes and appendices by R. Hermann, Math. Sci. Press, Brookline, MA, 1983, xiv+506 pp. (MR 85m:53001; Zbl. 603.53001.)

[5]Cartan, É., *Rimanova geometriya v ortogonal'nom repere* (Russian) [Riemannian geometry in an orthogonal coordinate system]; From lectures delivered by Élie Cartan at the Sorbonne in 1926–1927, translated and edited by S. P. Finikov, Izdat. Moskov. Univ., Moscow, 1960, 307 pp. (MR **8**, 602; Zbl. **99**, p. 372).

into Chinese.[6], and in 1998, when the original print run of 3,500 Russian copies of the book had long sold out, it was reprinted in Russia.[7]

Élie Cartan's book *Riemannian Geometry in an Orthogonal Frame* (*RGOF*) was available neither in English nor in French. This was the reason I suggested that World Scientific Publishing Co. translate *RGOF* into English. The company accepted my proposal.

Now the book (*RGOF*) is being presented for the first time in English. (It is not yet available in French!)

Note that I did not translate the last part of the Russian edition since this part is available in English (and French) as a part of the second edition of Cartan's book on the geometry of Riemannian manifolds (see the third and fourth footnotes).

I would like to note that *RGOF* contains many innovations. For example, the notion of intrinsic normal differentiation and of the Gaussian torsion of a submanifold in a Euclidean n-space or in a space of constant curvature, as well as an affine connection defined in a normal fiber bundle of a submanifold, were introduced for the first time in Cartan's lectures at the Sorbonne in 1926–1927. This is why *RGOF* was so often cited after its publication in 1960 (mostly in the papers of Russian geometers).

In my translation I have tried to follow the original text as closely as possible. I have made a few comments in the footnotes, marked as *Translator's note*. All references are given either in the text or the footnotes. I have verified all but two of the references and have appended the location of the corresponding reviews in: *Jahrbuch für Fortschritte der Mathematik* (JFM), *Zentralblatt für Mathematik* (Zbl), and *Mathematical Reviews* (MR). The two exceptions are signaled by (?).

I am deeply grateful to Shiing-Shen Chern for supporting the idea of translating this book of É. Cartan and kindly offering to write the Foreword to the translation. I am indebted to two people of World Scientific Publishing Co.: Sen Hu, Senior Editor, for his patience, interest, and assistance and Ye Qiang for an accurate and constructive copy-editing of the translation. I would also like to thank Maks A. Akivis and Eugene V. Ferapontov for valuable discussions in regard to some parts of my translation, Bruce Bukiet, who helped me to improve the style, and especially Richard W. Sharpe, who read my translation entirely and acted as its informal editor.

Vladislav V. Goldberg
Livingston, New Jersey, U.S.A.
March 2001

---

[6]Cartan, É, *Li-man chi-ho-hs—eh cheng chiao piao chia fa* (Chinese) [Riemannian geometry in an orthogonal coordinate system], translated from Russian by Chiang Li-fu, P'an Hsiao-jui, Huang Shu-t'ang, Ho Shao-hui, Yang Kan and Chou Ts'o-ling, Science Press, Peking, 1964, xii+267 pp. (MR **33** #3229.)

[7]Cartan, É., *Rimanova geometriya v ortogonal'nom repere*, reprint of the 1960 original, Platon, Volgograd, 1998.

Preface to the Russian Edition

Élie Cartan's well-known book on the geometry of Riemannian manifolds (1928) is based on lectures he delivered at the Sorbonne in 1925–1926. This course differed only slightly from the traditional exposition. In it Cartan emphasized the *geometry* of Riemannian manifolds, and tried "as far as possible to avoid very formal computations in which an orgy of indices hides a geometric picture which is often very simple."

In the second French edition of the book (1948) Cartan added a few new sections without altering the book's general construction. It is worth noting that in it Cartan gives a modern introduction to the notion of a manifold and, in particular, of a Riemannian manifold. He also considers the pseudo-Euclidean Riemannian manifolds in the spirit of global geometry, and in a masterly fashion, employs a Euclidean osculating space that allows an almost automatic transfer of geometric properties of curves in Euclidean space to those in a Riemannian manifold.

In his Sorbonne lectures of 1926–1927, Cartan introduces, at the very beginning, the method of exterior forms and uses orthogonal frames throughout.

In these lectures, Cartan's purpose is to *teach* his students. To this end, he first teaches them the new method of exterior forms, and then uses it to solve a series of intriguing problems in Euclidean space.

Mindful of the fact that Sorbonne's students must pass oral and written exams and that the written exam requires a solution to a problem posed by a professor, Cartan includes non-Euclidean geometries, a series of variational problems on geodesics, and solves many problems on the geometry of embedded submanifolds. All of these topics are cast in a remarkably simple form. However, they are entirely absent from both editions of the book.

This omission fascinated me as I translated the lectures of 1926–1927, devoted to the method of exterior forms, for my students. I later prepared these lectures for publication under the title *Riemannian geometry in an orthogonal frame*. Cartan's method of moving frames can also be applied successfully to non-orthogonal frames, or indeed to the frames of any transformation group, and the reader will find examples of this in this book. On the other hand, he will not find the theory of Riemannian manifolds in an orthogonal frame in any other book.

Although I consider this theory to be the main aim of the book, I could not refrain from including a series of articles from the 1925–1926 lectures that were not included in the Russian edition. These articles differ, even in style, from the lectures. I regret that I could not include all these articles in their entirety.

S. P. Finikov
Moscow, U.S.S.R.
March 1960

Contents

PRELIMINARIES

Chapter 1

Method of Moving Frames

1. Components of an infinitesimal displacement. Consider an ordinary three-dimensional Euclidean space. Take a right rectangular trihedron $\mathcal{T}$ at an arbitrary point M of this space. The set of all orthogonal trihedrons of the space depends on six parameters (three coordinates of the vertex and three Euler angles defining a rotation of a trihedron with respect to a fixed right orthogonal trihedron with the vertex O):

$$u_1, u_2, \ldots, u_6.$$

The trihedron $\mathcal{T}$ can be defined by the position vector $\mathbf{M} = \overrightarrow{OM}$ of its vertex and by the three mutually orthogonal unit vectors of its axes:

$$\mathbf{I}_1, \mathbf{I}_2, \mathbf{I}_3.$$

Define an infinitesimally close[1] trihedron $\mathcal{T}'$ that is associated with a point M'. The location of the new trihedron $\mathcal{T}'$ with respect to the trihedron $\mathcal{T}$ is determined by the increments of the position vector of its vertex,

$$\mathbf{M}' - \mathbf{M} = d\mathbf{M},$$

and unit vectors of its axes,

$$\mathbf{I}'_k - \mathbf{I}_k = d\mathbf{I}_k, \quad k = 1, 2, 3.$$

In the above formulas we denote the vectors defining an infinitesimally close trihedron by $\mathbf{M}'$ and $\mathbf{I}'_k$. If we project the differentials $d\mathbf{M}$ and $d\mathbf{I}_k$ onto the axes $\mathbf{I}_k$ of the trihedron $\mathcal{T}$, then we obtain the equations of an infinitesimal

---

[1] The coordinates $u_1 + du_1, \ldots, u_6 + du_6$ of its parameters are different by differentials $du_1, \ldots, du_6$ from the parameters $u_1, \ldots, u_6$ of the trihedron $\mathcal{T}$ associated with the point M.

(differential) displacement of the trihedron $\mathcal{T}$:

$$\begin{cases} d\mathbf{M} = \omega^1 \mathbf{I}_1 + \omega^2 \mathbf{I}_2 + \omega^3 \mathbf{I}_3, \\ d\mathbf{I}_1 = \omega_1^1 \mathbf{I}_1 + \omega_1^2 \mathbf{I}_2 + \omega_1^3 \mathbf{I}_3, \\ d\mathbf{I}_2 = \omega_2^1 \mathbf{I}_1 + \omega_2^2 \mathbf{I}_2 + \omega_2^3 \mathbf{I}_3, \\ d\mathbf{I}_3 = \omega_3^1 \mathbf{I}_1 + \omega_3^2 \mathbf{I}_2 + \omega_3^3 \mathbf{I}_3, \end{cases} \tag{1.1}$$

or briefly,

$$d\mathbf{M} = \omega^i \mathbf{I}_i, \quad d\mathbf{I}_i = \omega_i^j \mathbf{I}_j, \quad i, j = 1, 2, 3.$$

Here and in what follows, the summation symbol is omitted in sums when the sum index occurs twice, once up and once down. Note also that in the above formulas the quantities ω^i, $i = 1, 2, 3$, are the projections of the differential $d\mathbf{M}$ on the axes I_k. Hence they are infinitesimals and may be expressed linearly in terms of the differentials du^α of the independent variables u^α, $\alpha = 1, \dots, 6$. Thus they are linear differential forms, and the same is true for the quantities ω_i^j:

$$\begin{cases} \omega^i = \Gamma_\alpha^i du^\alpha, \\ \omega_i^j = \Gamma_{i\alpha}^j du^\alpha. \end{cases} \tag{1.2}$$

2. Relations among 1-forms of an orthonormal frame. The linear forms ω^i and ω_i^j cannot be given arbitrarily. First of all, they are connected by relations imposed by the fact that the vectors $\mathbf{I}_k$ of the trihedron $\mathcal{T}$ are orthonormal. This condition is expressed by the following formulas:

$$\begin{cases} \mathbf{I}_1^2 = 1, \qquad \mathbf{I}_2^2 = 1, \qquad \mathbf{I}_3^2 = 1, \\ \mathbf{I}_1 \cdot \mathbf{I}_2 = 0, \quad \mathbf{I}_2 \cdot \mathbf{I}_3 = 0, \quad \mathbf{I}_3 \cdot \mathbf{I}_1 = 0. \end{cases} \tag{1.3}$$

Differentiating these equations, we find that

$$\mathbf{I}_i \cdot d\mathbf{I}_i = 0 \ (i \text{ not summed}), \quad \mathbf{I}_j \cdot d\mathbf{I}_i + \mathbf{I}_i \cdot d\mathbf{I}_j = 0, \quad i, j = 1, 2, 3.$$

Substituting the values of the differentials $d\mathbf{I}_i$ taken from equations (1.1) into the above equations and taking into account (1.3), we find that

$$\omega_i^i = 0 \ (i \text{ not summed}), \quad \omega_i^j + \omega_j^i = 0, \quad i \neq j. \tag{1.4}$$

Thus there are only three independent 1-forms among the forms ω_i^j, namely the forms ω_2^3, ω_3^1, and ω_1^2. These 1-forms can be called the *components of rotation of the trihedron*. The components ω^i are called the *components of translation of the trihedron*.

Later, we will see that the forms ω^i and ω_i^j satisfy some additional differential relations.

3. Finding the components of a given family of trihedrons. If a motion of a trihedron is given, i.e., if one knows the coordinates of the position vector $\mathbf{M}$ and the unit axis vectors $\mathbf{I}_k$, then the components ω^i of the translation and the components ω_i^j of the rotation of the trihedron can be easily calculated by means of equations (1.1).

Suppose that the vectors $\mathbf{M}$ and $\mathbf{I}_k$ are given by their coordinates,

$$\mathbf{M}(x, y, z), \quad \mathbf{I}_i(\alpha_i, \beta_i, \gamma_i).$$

Multiplying the first equation of (1.1) by the vector $\mathbf{I}_1$, we find that

$$\mathbf{I}_1 \cdot d\mathbf{M} = \omega^1 \mathbf{I}_1 \cdot \mathbf{I}_1 + \omega^2 \mathbf{I}_1 \cdot \mathbf{I}_2 + \omega^3 \mathbf{I}_1 \cdot \mathbf{I}_3,$$

or, by equations (1.3),

$$\omega^1 = \mathbf{I}_1 \cdot d\mathbf{M}, \tag{1.5}$$

or

$$\omega^1 = \mathbf{I}_1(\alpha_1, \beta_1, \gamma_1) \cdot d\mathbf{M}(dx, dy, dz).$$

By the scalar product formula, the last equation implies that

$$\omega^1 = \alpha_1 dx + \beta_1 dy + \gamma_1 dz.$$

In general, we have

$$\omega^i = \alpha_i dx + \beta_i dy + \gamma_i dz. \tag{1.6}$$

Similarly, for example, we have

$$\omega_2^3 = \mathbf{I}_3 \cdot d\mathbf{I}_2 = \alpha_3 d\alpha_2 + \beta_3 d\beta_2 + \gamma_3 d\gamma_2 \tag{1.7}$$

and in general,

$$\omega_i^j = \alpha_j d\alpha_i + \beta_j d\beta_i + \gamma_i d\gamma_j. \tag{1.8}$$

Theorem 1. *For any choice of six parameters u^α, $\alpha = 1, \ldots, 6$, the six forms ω^i, $\omega_i^j = -\omega_j^i$ are linearly independent.*

Proof. To prove this theorem, it is sufficient to note that not only does the vanishing of the differentials du^α, $\alpha = 1, \ldots, 6$, make the differentials $d\mathbf{M}$ and $d\mathbf{I}_i$ vanish, and as a result, the forms ω^i, ω_i^j, vanish, but also conversely: by equations (1.1), if $\omega^i = 0$ and $\omega_i^j = 0$, then $d\mathbf{M} = 0$, $d\mathbf{I}_i = 0$, the trihedron is fixed, and all parameters u^α are constant. $\square$

4. Moving frames. In a similar way, with a point M of the Euclidean space, instead of an orthogonal trihedron (more precisely, an *orthonormal* trihedron, with unit axial vectors) one can associate an arbitrary oblique (nondegenerate) trihedron. The latter has variable angles and lengths of basis vectors with the only requirement that these basis vectors are linearly independent (i.e., they do not lie in a plane).

More precisely, to any point with position vector $\mathbf{M}$, we associate three arbitrary vectors $\mathbf{e}_i$, $i = 1, 2, 3$, with nonvanishing triple scalar product,

$$(\mathbf{e}_1, \mathbf{e}_2, \mathbf{e}_3) \neq 0,$$

i.e., the determinant composed of the coordinates of these vectors with respect to a fixed coordinate system does not vanish.

The figure formed by the point M and three vectors $\mathbf{e}_i$ (i.e., a moving system of oblique coordinates), is called a *frame* or a *moving frame*. The vectors $\mathbf{M}$ and $\mathbf{e}_i$ can be considered as functions of $3 + 3 \cdot 3 = 12$ arbitrary parameters u^α, $\alpha = 1, \dots, 12$ (their coordinates with respect to a fixed coordinate system). Then the differentials $d\mathbf{M}$ and $d\mathbf{e}_i$ are homogeneous linear functions of the differentials du^α of the independent variables u^α. The vectors $d\mathbf{M}$ and $d\mathbf{e}_i$ decompose with respect to the axes of our moving system, i.e., with respect to the triple of linearly independent vectors $\mathbf{e}_1, \mathbf{e}_2, \mathbf{e}_3$, and we obtain the following system of equations which is similar to the system of equations (1.1):

$$\begin{cases} d\mathbf{M} = \omega^1 \mathbf{e}_1 + \omega^2 \mathbf{e}_2 + \omega^3 \mathbf{e}_3, \\ d\mathbf{e}_i = \omega_i^1 \mathbf{I}_1 + \omega_i^2 \mathbf{I}_2 + \omega_i^3 \mathbf{I}_3, \end{cases} \tag{1.9}$$

where, as before, ω^i and ω_i^j are linear forms with respect to the differentials du^α of the independent variables u^α. However, formulas (1.3) are not valid, and as a result, equations (1.4) are also not valid.

5. Line element of the space. Introduce the notation

$$\mathbf{e}_i \cdot \mathbf{e}_j = g_{ij}, \quad g_{ij} = g_{ji}, \tag{1.10}$$

for the scalar products of the basis vectors. According to the geometric meaning of the scalar product, the quantity g_{ij} is equal to the product of the lengths of the vectors $\mathbf{e}_i$ and $\mathbf{e}_j$ multiplied by the cosine of the angle between them. On the other hand, squaring both parts of the first equation of (1.9), we obtain the square of the line element of the space:

$$ds^2 = (d\mathbf{M})^2.$$

Using the rule of squaring a polynomial (the sum of squares and double products of all mixed terms of a polynomial), in notations (1.10), we obtain

$$\begin{aligned} ds^2 &= (\mathbf{e}_1)^2 (\omega^1)^2 + \dots + 2\mathbf{e}_1 \cdot \mathbf{e}_2 \omega^1 \omega^2 + \dots \\ &= g_{11}(\omega^1)^2 + g_{22}(\omega^2)^2 + g_{33}(\omega^3)^2 \\ &\quad + 2g_{12}\omega^1\omega^2 + 2g_{23}\omega^2\omega^3 + 2g_{31}\omega^3\omega^1. \end{aligned} \tag{1.11}$$

This can be written in a shorter form:

$$ds^2 = g_{ij}\omega^i\omega^j. \tag{1.12}$$

This formula is valid also for an n-dimensional Euclidean space when the indices i and j run over the values $1, 2, \ldots, n$.

In formula (1.12), there are two pairs of identical indices, i and j, and there are two independent summations. For each pair of distinct indices, for example, 1 and 2, there are two terms: one with $i = 1$ and $j = 2$, and another with $i = 2$ and $j = 1$. By the symmetry of the coefficients g_{ij} (see (1.10)), these two terms are equal. As a result, in the expanded sum (1.11), all terms with different indices $i, j\, (i \neq j)$ enter with the coefficient 2. The scalar product of two vectors

$$\mathbf{X} = X^i \mathbf{e}_i, \quad \mathbf{Y} = Y^i \mathbf{e}_i$$

can be computed in a similar way. Applying the scalar product rules, we find that

$$\mathbf{X} \cdot \mathbf{Y} = \mathbf{e}_i \cdot \mathbf{e}_j X^i Y^j = g_{ij} X^i Y^j. \tag{1.13}$$

It follows that the cosine of the angle φ between two vectors $\mathbf{X}$ and $\mathbf{Y}$ is

$$\cos \varphi = \frac{g_{ij} X^i Y^j}{\sqrt{g_{ij} X^i X^j} \sqrt{g_{ij} Y^i Y^j}}. \tag{1.14}$$

6. Contravariant and covariant components. Multiply a vector

$$\mathbf{X} = X^k \mathbf{e}_k \tag{1.15}$$

by the vector $\mathbf{e}_i$. If we denote the product $\mathbf{X} \cdot \mathbf{e}_i$ by X_i and apply formula (1.10), then we find that

$$X_i = g_{ik} X^k. \tag{1.16}$$

The components X^k are called the *contravariant* components of a vector $\mathbf{X}$, and the components X_k are called the *covariant* components of a vector $\mathbf{X}$. The contravariant components are denoted by upper indices, and the covariant components are denoted by lower indices.

Multiplying both parts of equations (1.16) by X^i and summing with respect to the index i, we obtain

$$X_i X^i = g_{ik} X^i X^k = \mathbf{X}^2, \tag{1.17}$$

and similarly,

$$\mathbf{X} \cdot \mathbf{Y} = g_{ik} X^i Y^k = X^i Y_i. \tag{1.18}$$

In order to pass from covariant components to contravariant components, one must solve the system of n equations $(i = 1, 2, \ldots, n)$ (1.16) with n unknowns X^k, $k = 1, \ldots, n$. By the well-known algebraic formula, a solution has the form

$$X^k = g^{ik} X_i, \tag{1.19}$$

where g^{ik} is the reduced cofactor of the entry g_{ik} in the determinant

$$g = \det(g_{ij}) \neq 0,$$

i.e, it is the cofactor of the entry g_{ik} divided by the determinant g. All these can be also applied to the infinitesimal components ω^i or ω_i.

For an orthonormal frame, we have

$$g_{ij} = \delta_{ij} = \begin{cases} 0, & \text{if } i \neq j, \\ 1, & \text{if } i = j. \end{cases}$$

From equations (1.16), it follows that

$$X_i = X^i,$$

i.e., the covariant components of a vector are equal to its contravariant components. When we use orthonormal frames, we write without distinction $\omega^i = \omega_i$ and $\omega_i^j = \omega_{ij}$, and as before, we assume that repeated indices, one up and one down, imply the summation.[2]

7. Infinitesimal affine transformations of a frame. Note also that for an oblique frame, we can no longer say that equations (1.9) determine a frame motion since our trihedrons are not congruent. However, it is not difficult to check that under the passage from one coordinate trihedron to another one, the transformation of coordinates x^1, x^2, x^3 is determined by a linear substitution. In fact, if we denote by a_i^j the cosines of the angles between the old and new axes, by x^i and $\overline{x}^i$ the old and new coordinates of an arbitrary point, and by x_0^i the coordinates of the new origin with respect to the old coordinate system, then we obtain the following expressions of the old coordinates:

$$x^i = x_0^i + a_1^i \overline{x}^1 + a_2^i \overline{x}^2 + a_3^i \overline{x}^{3'}.$$

Since a general linear substitution defines an affine transformation, *equations (1.9) define infinitesimal affine transformations of a frame.*

---

[2]Cartan will use this many times in the following chapters, sometimes in order to indicate that there is no summation and sometimes to match a location of indices on the left-hand and right-hand sides of an equation. This is also the reason that Cartan uses both terms "the vector" and "the form" for the quantities like $\omega^i = \omega_i$. (*Translator's note.*)

Chapter 2

The Theory of Pfaffian Forms

We return to an orthonormal frame. We saw that if a frame motion is given, i.e., if the coordinates of the position vector $\mathbf{M}$ and the unit axial vectors $\mathbf{I}_i$ are given as functions of the parameters u^α, then we can find the components ω^i and ω_i^j of an infinitesimal displacement of a frame.

A problem arises: could six arbitrarily given forms ω^i and ω_i^j which are linear combinations of du^α represent a displacement of a trihedron in an ordinary Euclidean space?

In order to answer this question, we make a digression into analysis and consider properties of Pfaffian forms.

8. Differentiation in a given direction. Suppose that a Pfaffian form

$$\omega = a_1 dx^1 + a_2 dx^2 + \ldots + a_n dx^n$$

is given. Usually the differentials $dx^1, \ldots, dx^n$ of the independent variables $x^1, \ldots, x^n$ are considered as new independent variables. However, we can also take them to be given functions of the independent variables,

$$dx^i = \xi^i(x^1, \ldots, x^n)$$

(a *directed displacement* with a definite symbol of differentiation). Consider two symbols of differentiation d and δ and two series of differentials

$$d\delta x^1, d\delta x^2, \ldots, d\delta x^n;$$

$$\delta dx^1, \delta dx^2, \ldots, \delta dx^n,$$

which in general are distinct.

We assume that these two symbols of differentiation commute,

$$d\,\delta x^i = \delta\,dx^i. \tag{2.1}$$

Theorem 1. *If the symbols of differentiation d and δ commute for independent variables, then they also commute for functions.*

Proof. In fact, consider a function

$$f(x^1, \ldots, x^n).$$

Then

$$df = \frac{\partial f}{\partial x^i} dx^i, \quad \delta f = \frac{\partial f}{\partial x^i} \delta x^i.$$

The second differentiation gives

$$\delta\, df = \frac{\partial^2 f}{\partial x^i \partial x^k} dx^i \delta x^k + \frac{\partial f}{\partial x^i} \delta\, dx^i,$$

$$d\, \delta f = \frac{\partial^2 f}{\partial x^i \partial x^k} \delta x^i dx^k + \frac{\partial f}{\partial x^i} d\, \delta x^i. \qquad (2.2)$$

By condition (2.1), the second terms of these two expressions are the same. As for the first terms, if we change the indices of summation i for k and k for i in the first term of the expression for $d\,\delta f$, then we obtain the sum

$$\frac{\partial^2 f}{\partial x^k \partial x^i} \delta x^k dx^i,$$

which differs from a similar sum in the expression of $\delta\, df$ only by the order of differentiation in the second partial derivatives. Since in the mixed partial derivatives the result of differentiation does not depend on the order of differentiation, the first terms on the right-hand sides of expressions (2.2) coincide, and we have

$$\delta\, df = d\, \delta f. \qquad (2.3)$$

Corollary 2. *The property of commutativity for two symbols of differentiation d and δ does not depend of the choice of independent variables.*

The situation is different for the linear (Pfaffian) forms.

9. Bilinear covariant of Frobenius. Consider a Pfaffian form

$$\omega(d) = a_i dx^i,$$

and compose an expression

$$d\omega(\delta) - \delta\, d\omega(d).$$

Differentiating $\omega(d)$ by means of differentiation δ, we obtain

$$\delta\, \omega(d) = \delta a_i\, dx^i + a_i\, \delta\, dx^i,$$

and similarly,
$$d\omega(\delta) = da_i\,\delta x^i + a_i\,d\,\delta x^i.$$

Subtracting the first expression from the second, we find that
$$d\omega(\delta) - \delta\,\omega(d) = da_i\,\delta x^i - \delta a_i\,dx^i = \frac{\partial a_i}{\partial x^k}\,dx^k\delta x^i - \frac{\partial a_i}{\partial x^k}\,\delta x^k dx^i.$$

Change the summation index i for k and k for i in the last sum. As a result, we get
$$\frac{\partial a_i}{\partial x^k}\,dx^k\delta x^i - \frac{\partial a_k}{\partial x^i}\,\delta x^i dx^k = \left(\frac{\partial a_i}{\partial x^k} - \frac{\partial a_k}{\partial x^i}\right)dx^k\delta x^i.$$

If we sum over combinations of i and k, then for each pair of indices i and k, we have two terms
$$\left(\frac{\partial a_i}{\partial x^k} - \frac{\partial a_k}{\partial x^i}\right)dx^k\delta x^i + \left(\frac{\partial a_k}{\partial x^i} - \frac{\partial a_i}{\partial x^k}\right)dx^i\delta x^k$$
$$= \left(\frac{\partial a_i}{\partial x^k} - \frac{\partial a_k}{\partial x^i}\right)(dx^k\delta x^i - dx^i\delta x^k).$$

If we denote the summation over combinations of i and k by (i,k), then we find that
$$d\omega(\delta) - \delta\,\omega(d) = \sum_{(i,k}\left(\frac{\partial a_i}{\partial x^k} - \frac{\partial a_k}{\partial x^i}\right)(dx^k\delta x^i - dx^i\delta x^k). \qquad (2.4)$$

The form on the right-hand side of equation (2.4) is linear with respect to the differentials $dx^1,\ldots,dx^n$ as well as with respect to the differentials $\delta x^1,\ldots,\delta x^n$. This form is invariant with respect to a change of independent variables, i.e., under a change of the variables x^i in both parts of (2.4), the equality will be preserved. Expression (2.4) is called the *bilinear covariant of Frobenius* (it is named after the scientist who introduced such forms).

Theorem 3. *If the bilinear covariant of Frobenius of a Pfaffian form ω vanishes,*
$$d\omega(\delta) - \delta\,\omega(d) = 0, \qquad (2.5)$$

then the form ω is a total differential.

Proof. In fact, the vanishing of form (2.4) for arbitrary differentials dx^i and δx^i implies the vanishing of the coefficients on the right-hand side of (2.4):
$$\frac{\partial a_i}{\partial x^k} - \frac{\partial a_k}{\partial x^i} = 0.$$

It is well-known that the last equation is necessary and sufficient for the form
$$\omega = a_i\,dx^i$$

to be a total differential, $\omega = d\varphi$. $\qquad\qquad\square$

Theorem 4. *If the bilinear covariant of a Pfaffian form ω vanishes, then the integral*

$$u = \int_{M_0}^{M_1} \omega(d)$$

between two points $M_0(x_0^1, \ldots, x_0^n)$ and $M_1(x_1^1, \ldots, x_1^n)$ does not depend on the choice of a path of integration, i.e., it is path independent. It is assumed that paths of integration connecting two points belong to a continuous family of curves.

Proof. This theorem is equivalent to the previous one. However, we give an independent proof of this theorem which can be also considered as the proof of the previous theorem.

Consider a continuous family of paths connecting the points M_0 and M_1:

$$x^i = x^i(t, \alpha), \quad i = 1, \ldots n,$$

where α is a parameter of the lines of the family, and t is the parameter of a point on a curve α. The parameter t is chosen in such a way that $t = 0$ corresponds to the point M_0, and $t = 1$ corresponds to the point M_1. Hence

$$x^i(0, \alpha) = x_0^i \quad \text{and} \quad x^i(1, \alpha) = x_1^i$$

are independent of the choice of α.

Define two symbols of differentiation:

$$d = \frac{\partial}{\partial t}, \quad \delta = \frac{\partial}{\partial \alpha}.$$

It is evident that they commute, since we have

$$\delta\, df = \frac{\partial}{\partial \alpha}\left(\frac{\partial f}{\partial t}\right) = \frac{\partial^2 f}{\partial \alpha\, \partial t}, \quad d\,\delta f = \frac{\partial}{\partial t}\left(\frac{\partial f}{\partial \alpha}\right) = \frac{\partial^2 f}{\partial t\, \partial \alpha}.$$

By the theorem hypothesis, the bilinear covariant of Frobenius of the form ω vanishes,

$$d\omega(\delta) - \delta\, d\omega(d) = 0.$$

We set

$$u(t, \alpha) = \int_0^t \omega(d). \tag{2.6}$$

It follows that

$$\omega(d) = \frac{\partial u}{\partial t} = du, \quad \delta\, \omega(d) = \delta\, du,$$

and equation (2.5) takes the form

$$d\omega(\delta) - \delta\, du = 0,$$

or by (2.1), the form
$$d[\omega(\delta) - \delta u] = 0,$$
for any t and α. Thus
$$\frac{\partial}{\partial t}[\omega(\delta) - \delta u] = 0.$$
Since the derivative with respect to t vanishes, the difference $\omega(\delta) - \delta u$ does not depend on t. But at $t = 0$, i.e., at the point M_0, all paths come together, and from formula (2.6) we see that the initial value $u = 0$. Hence
$$\omega(\delta) = a_i \frac{\partial x^i}{\partial \alpha} = 0, \quad u = 0,$$
and for each t, we have
$$\omega(\delta) - \delta u = 0.$$
At the point M_1, i.e., for $t = 1$, all paths come together again, and as a result, we have
$$\omega(\delta) = 0.$$
Thus, for $t = 1$,
$$\delta u = 0,$$
i.e., for any integration path of a continuous family of curves, integral (2.6) has one and the same value.   □

A bilinear covariant depends on two series of differentials dx^i and δx^i and changes the sign under their permutation. Bilinear algebraic forms of two series of variables u^i and v^k, that change their signs under the permutation of u^i and v^k, are called *skew-symmetric*.

10. Skew-symmetric bilinear forms.

1°. *The coefficients of skew-symmetric forms are skew-symmetric themselves, i.e., they change their signs under permutation of their indices.*

Let
$$F = a_{ij}u^i v^j \tag{2.7}$$
be a skew-symmetric form. Then after permutation of variables u and v, it becomes
$$F = -a_{ij}v^i u^j.$$
Changing the index of summation i for j and j for i in the last equation, we find that
$$F = -a_{ji}v^j u^i. \tag{2.8}$$
Comparing (2.7) with (2.8), we obtain
$$a_{ij}u^i v^j = -a_{ji}v^j u^i,$$
and as a result, we have
$$a_{ij} = -a_{ji}, \quad a_{ii} = 0 \ (i \text{ not summed}).$$

2°. *A skew-symmetric form remains skew-symmetric under simultaneous (with the same coefficients) linear substitution of both series of variables.*

Consider a substitution

$$U^i = A^i_k u^k, \quad V^i = A^i_k v^k.$$

It is obvious that, under this substitution, the form F will be transferred into a bilinear form Φ,

$$F(u, v) = \Phi(U, V).$$

In fact, it is clear that if we change u^k for v^k, then U^i will be changed for V^i, and the form Φ will change its sign simultaneously with the form F. Hence

$$\Phi(U, V) = -\Phi(V, U).$$

11. Exterior quadratic forms. If in formula (2.7) we turn to the summation over permutation of the indices i and j and apply the fact that the coefficients a_{ij} are skew-symmetric, $a_{ij} = -a_{ji}$, then we obtain the following expression for the form F:

$$F = \sum_{(i,j)} a_{ij}(u^i v^j - u^j v^i) = \sum_{(i,j)} a_{ij} \begin{vmatrix} u^i & u^j \\ v^i & v^j \end{vmatrix}.$$

Introduce a new notation

$$u^i \wedge u^j = \begin{vmatrix} u^i & u^j \\ v^i & v^j \end{vmatrix}. \tag{2.9}$$

We call the product $u^i \wedge u^j$ the *exterior product*. The determinant on the right-hand side of (2.9) is called a *value of the exterior product*. In order to obtain a value of the exterior product, it is necessary to give two series of values of the basis elements, u^i, u^j and v^i, v^j.

All operations on exterior products are subject to the general rule: *any transformation of an expression with exterior products is valid if the values of this expression before and after the transformation are the same for any choice of basis variables.*

For example, if we have two linear forms,

$$f_1 = a_i u^i, \quad f_2 = b_j u^j,$$

then

$$f_1 \wedge f_2 = \begin{vmatrix} f_1(u) & f_2(u) \\ f_1(v) & f_2(v) \end{vmatrix} = \begin{vmatrix} a_i u^i & b_k u^k \\ a_i v^i & b_k v^k \end{vmatrix} = (a_i u^i)(b_k v^k) - (a_i v^i)(b_k u^k)$$

$$= a_i b_k (u^i v^k - v^i u^k) = a_i b_k \begin{vmatrix} u^i & u^k \\ v^i & v^k \end{vmatrix} = a_i b_k \, u^i \wedge u^k$$

$$= \sum_{(i,k)} (a_i b_k - a_k b_i) \, u^i \wedge u^k.$$

Corollary 5. 1. *The exterior product is anticommutative.*

2. *The exterior product of equal factors or proportional factors is equal to 0:*

$$f \wedge f = 0, \quad f \wedge (cf) = 0.$$

3. *A scalar factor can be taken across of an exterior product.*

12. Converse theorems. Cartan's Lemma.

Theorem 6. *The equation*

$$f \wedge \varphi = 0, \tag{2.10}$$

where f is a given linear form, holds if and only if the form φ is proportional to f:

$$\varphi = cf.$$

Proof. The sufficiency follows from the second part of Corollary 5. To prove the necessity, we replace the basis $u^1, u^2, \ldots, u^n$ by a new basis consisting of the given form f and $n-1$ arbitrary forms $f_1, f_2, \ldots, f_{n-1}$ provided that $f, f_1, f_2, \ldots, f_{n-1}$ are linearly independent, i.e., the determinant composed of their coefficients with respect to the old basis is not zero. For example, if the decomposition of the form f with respect to the old basis contains u^n, then we can set

$$f_i = u^i, \quad i = 1, 2, \ldots, n-1.$$

In the new basis, the form φ has a representation

$$\varphi = cf + c^1 f_1 + \ldots + c^{n-1} f_{n-1},$$

and equation (2.10) takes the form

$$cf \wedge f + c^1 f \wedge f_1 + \ldots + c^{n-1} f \wedge f_{n-1} = 0.$$

By the second part of Corollary 5, the first term of the last equation vanishes. All other terms of this equation are linearly independent as exterior products of different basis elements. Therefore, the coefficients $c^1, \ldots, c^{n-1}$ must vanish,

$$c^1 = \ldots = c^{n-1} = 0,$$

and as a result, we have

$$\varphi = cf.$$

Theorem 7 (Cartan's Lemma). *Suppose that $f^1, f^2, \ldots, f^p$ is a given set of linearly independent forms, (i.e., the rank of the matrix of their coefficients is equal to p). Then the equation*

$$f^1 \wedge \varphi_1 + f^2 \wedge \varphi_2 + \ldots f^p \wedge \varphi_p = 0, \tag{2.11}$$

holds if and only if the forms $\varphi_i, i = 1, \ldots, p$, are linear combinations of the forms $f^k, k = 1, \ldots, p$, with a symmetric matrix of coefficients:

$$\varphi_i = c_{ik} f^k, \quad c_{ik} = c_{ki}. \tag{2.12}$$

Proof. Enlarge the set of forms $f^1, f^2, \ldots, f^p$ by $n - p$ forms $f^{n+1}, \ldots, f^n$ in such a way that the forms $f^1, f^2, \ldots, f^n$ form a basis. Since the rank of the matrix of coefficients in the decompositions of the forms $f^k, k = 1, \ldots, p$, with respect to the basis $u^1, \ldots, u^n$ is equal to p, this coefficient matrix has at least one nonvanishing determinant of order p.

Suppose that this determinant is formed by the coefficients in the basis elements $u^1, \ldots, u^p$. Then we can take

$$f^\lambda = u^\lambda, \quad \lambda = p + 1, \ldots, n.$$

Suppose that the decomposition of the forms $\varphi_i, i = 1, \ldots, p$ with respect to the new basis is

$$\varphi_i = c_{ik} f^k + c_{i\lambda} f^\lambda, \quad k = 1, \ldots, p; \lambda = p + 1, \ldots, n.$$

Then

$$f^i \wedge \varphi = c_{ik} f^i \wedge f^k + c_{i\lambda} f^i \wedge f^\lambda,$$

and equation (2.11) takes the form

$$c_{ik} f^i \wedge f^k + c_{i\lambda} f^i \wedge f^\lambda = 0. \tag{2.13}$$

Each term of the second sum in (2.13) contains a new form f^λ which does not occur anywhere else in equation (2.13). Thus all the coefficients $c_{i\lambda}$ vanish,

$$c_{i\lambda} = 0.$$

As a result, *the forms φ decompose in terms of the forms $f^k, k = 1, \ldots, p$, only.* When we sum with respect to the combinations of i and k, there will be two similar (with the same forms f^i and f^k) terms in the first sum of (2.13):

$$c_{ik} f^i \wedge f^k + c_{ki} f^k \wedge f^i = (c_{ik} - c_{ki}) f^i \wedge f^k.$$

Now the equation

$$(c_{ik} - c_{ki})f^i \wedge f^k = 0$$

does not have similar terms. Thus, all the coefficients are equal to 0,

$$c_{ik} - c_{ki} = 0.$$

This leads to formulas (2.12). $\square$

Theorem 8. *Suppose that the basis elements u^i, $i = 1, \ldots, n$, are connected by the equation*

$$f = c_i u^i = 0. \tag{2.14}$$

Then the exterior quadratic form

$$F = \sum_{(i,k)} a_{ij} u^i \wedge u^j \tag{2.15}$$

vanishes by virtue of (2.14) if and only if F has the form

$$F = f \wedge \varphi, \tag{2.16}$$

where φ is an appropriately chosen linear form.

Proof. In fact, if we take a new basis $f, f^1, \ldots, f^{n-1}$, then equation (2.15) becomes

$$F = b_j f \wedge f^j + \sum_{(i,j)} b_{ij} f^i \wedge f^j, \quad i, j = 1, \ldots, n.$$

Taking into account condition (2.14) and $F = 0$, we find that

$$\sum_{(i,j)} b_{ij} f^i \wedge f^j, \quad \text{i.e.,} \quad b_{ij} = 0.$$

As a result, we obtain the following expression for the form F:

$$F = b_j f \wedge f^j = f \wedge (b_j f^j).$$

If in the last formula we set

$$\varphi = b_j f^j,$$

then we arrive at equation (2.16). $\square$

Theorem 9. *Suppose that the basis elements u^i, $i = 1, \ldots, n$, are connected by a system of equations*

$$f^1 = 0, \ldots, f^p = 0, \tag{2.17}$$

where f^i, $i = 1, \ldots, p$, are linearly independent forms. Then the exterior quadratic form F vanishes by virtue of (2.17) if and only if

$$F = f^1 \wedge \varphi_1 + f^2 \wedge \varphi_2 + \ldots + f^p \wedge \varphi_p. \tag{2.18}$$

Proof. Sufficiency is clear. We prove necessity. In fact, since the forms f^i, $i = 1, \ldots, p$, are linearly independent, we can enlarge them by the forms f^λ, $\lambda = p+1, \ldots, n$, to a basis. In the new basis, the form F has the following decomposition:

$$F = c_{ij} f^i \wedge f^j + c_{i\lambda} f^i \wedge f^\lambda + c_{\lambda\mu} f^\lambda \wedge f^\mu, \quad i, j = 1, \ldots, p; \; \lambda, \mu = p+1, \ldots, n.$$

Since F vanishes on the solutions of equations (2.17), the last equation gives

$$c_{\lambda\mu} f^\lambda \wedge f^\mu = 0.$$

Thus the expression for the form F becomes

$$F = f^i \wedge (c_{ij} f^j + c_{i\lambda} f^\lambda).$$

If in the last formula we set

$$\varphi_i = c_{ij} f^j + c_{i\lambda} f^\lambda,$$

then we arrive at equation (2.18). $\qquad\qquad\qquad\qquad\qquad\qquad\qquad\square$

13. Exterior differential. Just as a bilinear form can be represented by an exterior quadratic form, a bilinear covariant of Frobenius

$$d\omega(\delta) - \delta\omega(d) = (da_1 \, \delta x^1 - \delta a_1 \, dx^1) + \ldots + (da_p \, \delta x^p - \delta a_p \, dx^p)$$

of the form

$$\omega = a_1 \, dx^1 + \ldots + a_p \, dx^p$$

can be represented by the *exterior differential*

$$d\omega = da_1 \wedge dx^1 + \ldots + da_p \wedge dx^p. \tag{2.19}$$

Let us find *the rule of exterior differentiation of the product of a scalar function and a form.* Let m be a function of variables $x^1, \ldots, x^n$, and ω be a linear form,

$$\omega = a_i \, dx^i.$$

Compute the exterior differential of the product $m\omega$:

$$d(m\omega) = d[(ma_i)dx^i] = d(ma_i) \wedge dx^i.$$

But

$$d(ma_i) = m \, da_i + a_i \, dm.$$

Thus

$$d(m\omega) = md(a_i \wedge dx^i) + dm \wedge (a_i \, dx^i),$$

or

$$d(m\omega) = md\omega + dm \wedge \omega.$$

Chapter 3

Integration of Systems of Pfaffian Differential Equations

14. Integral manifold of a system. Consider a system of r Pfaffian equations (a Pfaffian system)

$$\theta_1 = 0, \ldots, \theta_r = 0, \tag{3.1}$$

where θ_α, $\alpha = 1, \ldots, r$, are r linear differential forms constructed by means of $n + r$ independent variables $x^1, x^2, \ldots, x^{n+r}$ and their differentials in such a way that the coefficients in differentials in representations of θ_α are of class C^1, i.e., these coefficients are continuous and have continuous first-order partial derivatives. Suppose that these forms are linearly independent. In other words, we assume that the rank of the matrix $\mathfrak{M}$ of their coefficients is equal to r.

Geometrically, this means that we consider variables x^i as coordinates of a point in a space of dimension $n + r$ and call a point A of this space a *generic point* if for this point, the rank of the matrix $\mathfrak{M}$ is precisely equal to r.

Consider in this space an n-dimensional smooth manifold V, i.e, a manifold such that the coordinates of any of its points can be given by smooth functions of n parameters. We say that V is an *integral manifold of system* (3.1), or that V provides a *solution of system* (3.1), if each of its tangent elements, defined by a point $A \in V$ and directing parameters $dx^1, dx^2, \ldots, dx^{n+r}$ of a tangent line to V at the point A, satisfies system (3.1). If a point A in question is a generic point of the space and the determinant formed by the last r columns of the matrix $\mathfrak{M}$ does not vanish at this point (this always can be arranged by changing, if necessary, the coordinate numbering), then in a neighborhood of the point A, system (3.1) can be solved with respect to the differentials $dx^{n+1}, \ldots, dx^{n+r}$. As a result, the variables $x^{n+1}, \ldots, x^{n+r}$, considered on the manifold V as functions of the parameters $x^1, \ldots, x^n$, admit continuous partial derivatives at the point A and all points sufficiently close to A.

For convenience of exposition, we set

$$x^{n+\alpha} = z^{\alpha}, \quad \alpha = 1, \ldots, r,$$

and leave only the values $1, \ldots, n$ for the indices i and j: $i, j = 1, \ldots, n$. Then we can write the forms θ_{α} as

$$\theta_{\alpha} = c_{\alpha\beta} dz^{\beta} + a_{\alpha i} dx^{i}, \quad \det(c_{\alpha\beta}) \neq 0, \quad \alpha, \beta = 1, \ldots, r, \tag{3.2}$$

and define the manifold V by means of r equations

$$z^{\alpha} = z^{\alpha}(x^1, x^2, \ldots, x^n), \quad \alpha = 1, \ldots, r. \tag{3.3}$$

Since the functions z^{α} admit continuous first-order partial derivatives, we can say that any point of the integral manifold V, which is a generic point of the space, is a *regular point* of V. This means a possibility of an analytic representation of the manifold V in the form of (3.3) exists.

15. Necessary condition of complete integrability.

Definition 1. A Pfaffian system (3.1) is said to be *completely integrable* if through any generic point of the space, there passes one integral manifold.

Of course, we require the existence of an integral manifold in a sufficiently small neighborhood of the generic point in question.

It is not difficult to establish a necessary condition for complete integrability. In fact, compute the exterior differentials of the forms θ_{α}: they are exterior quadratic forms built on the variables x^k, $k = 1, \ldots, n+r$, and their differentials dx^k.

In a neighborhood of a generic point A, for which the conditions discussed above hold, equations (3.2) can be solved with respect to dz^{α}, and the exterior quadratic forms $d\theta_{\alpha}$ are expressed in terms of the n differentials dx^i and the r linearly independent forms θ_{α}. As a result, we obtain

$$d\theta_{\alpha} = \frac{1}{2} A_{\alpha ij} dx^i \wedge dx^j + B^{\beta}_{\alpha i} dx^i \wedge \theta_{\beta} + \frac{1}{2} C^{\beta\gamma}_{\alpha} \theta_{\beta} \wedge \theta_{\gamma}, \tag{3.4}$$

where the summation indices run over the following values: $i, j = 1, \ldots, n; \beta, \gamma = 1, \ldots, r$, and

$$A_{\alpha ij} = -A_{\alpha ji}, \quad C^{\beta\gamma}_{\alpha} = -C^{\gamma\beta}_{\alpha}.$$

Now let V be an integral manifold passing through a generic point A. If one moves along this manifold, then the forms θ_{α} as well as their exterior differentials $d\theta_{\alpha}$ are identically equal to zero. So, the left-hand side and the last two sums on the right-hand side of equations (3.4) vanish. On the other hand, in a neighborhood of a point A, for which the analytic representation (3.3) takes place, the variables $x^1, \ldots, x^n$ remain independent. Thus the coefficients $A_{\alpha ij}$ in equations (3.4) are equal to zero. This implies that in this neighborhood, the equations of the form

$$d\theta_{\alpha} = \widetilde{\omega}^{\beta}_{\alpha} \wedge \theta_{\beta} \tag{3.5}$$

hold, where $\tilde{\omega}_\alpha^\beta$ are appropriately chosen forms with continuous functions as their coefficients. For example, we can set

$$\tilde{\omega}_\alpha^\beta = B_{\alpha i}^\beta dx^i + \frac{1}{2} C_\alpha^{\gamma\beta} \theta_\gamma.$$

Condition (3.4) can be viewed as a necessary condition of complete integrability of system (3.1). This condition can even be written in the simpler form

$$d\theta_\alpha \equiv 0 \quad (\mathrm{mod} \ \theta_1, \theta_2, \dots, \theta_r). \tag{3.6}$$

16. Necessary and sufficient condition of complete integrability of a system of Pfaffian equations. Now we will formulate a general theorem for completely integrable Pfaffian systems.

Theorem 2. *A Pfaffian system* (3.1) *is completely integrable if and only if in a neighborhood of a generic point of the space, the exterior differentials $d\theta_\alpha$ are congruent to zero modulo of $\theta_1, \theta_2, \dots, \theta_r$.*

This necessary and sufficient condition can be also expressed as follows: *the system of exterior differentials*

$$d\theta_1 = 0, \quad d\theta_2 = 0, \quad \dots, \quad d\theta_r = 0 \tag{3.7}$$

must be an algebraic consequence of system (3.1).

In fact, if $\theta_1 = 0, \theta_2 = 0, \dots, \theta_r = 0$, then equations (3.5) immediately imply system (3.7). On the other hand, it follows from Theorem 9 of Chap. 2 that the quadratic forms $d\theta_\alpha$, whose vanishing is an algebraic consequence of the system $\theta_1 = 0, \theta_2 = 0, \dots, \theta_r = 0$, have the structure (2.18), and this is equivalent to condition (3.5).

Proof. Now we prove Theorem 2. The necessity of condition (3.5) or (3.6) follows from our derivation of this condition. Its sufficiency follows from a few auxiliary theorems which we prove below.

It is convenient to consider a space of parameters with points (x^i).

Theorem 3. *System* (3.1) *admits at most one solution satisfying the initial conditions*

$$z^\alpha = z_0^\alpha \ \text{for} \ x^i = x_0^i; \quad z_0^\alpha, x^i = \text{const}. \tag{3.8}$$

Proof. Suppose that solution (3.3) exists. Then we can find its value at a point $M(x^i)$. In the space (x^i) choose a point $M_0(x_0^1, \dots, x_0^n)$ and assign the values $z_0^1, \dots, z_0^r$. Suppose that the point $M(x^i)$ along with the point M_0 lie in a simply connected domain where analytic representation (3.3) holds. We prove that at the point $M(x^i)$ the functions z^α have certain definite values. To this end, we join the point M_0 with the point M by a path passing inside the domain under consideration:

$$x^1 = f^1(t), \quad x^2 = f^2(t), \quad \dots, \quad x^n = f^n(t), \tag{3.9}$$

where all functions $f(t)$ are of class C^1, t is a parameter defining the location of a point on the line M_0M, and $t = t_0$ at the point M_0.

Substituting expressions (3.9) into system (3.1), dividing the equations obtained by the differential dt, and solving these equations with respect to the derivatives $\frac{dz^\alpha}{dt}$, we obtain a Cauchy system for unknown functions z^α:

$$\frac{dz^\alpha}{dt} = Z^\alpha(t, z^1, \ldots, z^r). \tag{3.10}$$

Our solution (3.3) must satisfy system (3.10), and since a Cauchy system admits only one solution with given initial conditions (3.8), this solution is unique. □

17. Path independence of the solution.

Theorem 4. *If condition* (3.5) *or* (3.6) *is satisfied, then in a neighborhood of a generic point M_0, for a given continuous family of paths joining the point M_0 with the point M_1, the value (z^α) is path independent.*

Proof. Consider a simply connected domain in the space of parameters (x^i). Consider also two points M_0 and M_1 in this domain that are sufficiently close so that when we integrate system (3.10) along any path of a continuous family of lines

$$x^1 = f^1(t, \alpha), \quad x^2 = f^2(t, \alpha), \quad \ldots, \quad x^n = f^n(t, \alpha), \tag{3.11}$$

joining the point M_0 with the point M_1, an interval, in which the Cauchy existence theorem for solution is valid, contains the point M_1 as its interior point.

In equations (3.11), the parameter t again determines a point on a line of the family M_0M_1 with the values $t = 0$ at the point M_0 and $t = 1$ at the point M_1. The parameter α in these equations determines a line of the family, and the functions $f^i(t, \alpha)$ are of class C^1. We integrate system (3.1) along every path of the family, emanating from the point M_0, and with the same initial values (3.8). We must prove that independently of the path of integration, we arrive at the point M_1 with the same values z^α.

Define two symbols of integrations:

1) along a path of integration

$$d = \frac{\partial}{\partial t};$$

2) for the passage from one path to another

$$\delta = \frac{\partial}{\partial \alpha}.$$

At the point M_0, i.e., for $t = 0$, all paths come together, and all unknown functions z^α take the same initial values $z^\alpha = z_0^\alpha$. Thus the following variations are equal to 0,

$$\delta x^i = 0, \quad \delta z^\alpha = 0,$$

and for $t = 0$, we have

$$\theta_\alpha(\delta) = 0, \quad \alpha = 1, \dots, r. \tag{3.12}$$

At the point M_1, i.e., for $t = 1$, all paths come together again, and as a result,

$$\delta x^i = 0.$$

It remains to prove that $\delta z^\alpha = 0$. From (3.2), for $t = 1$, we have

$$\theta_\alpha(\delta) = c_{\alpha 1} \frac{\partial z^1}{\partial \alpha} + c_{\alpha 2} \frac{\partial z^2}{\partial \alpha} + \dots + c_{\alpha r} \frac{\partial z^r}{\partial \alpha}. \tag{3.13}$$

We integrate system (3.1) at any point of a line $M_0 M_1$, i.e., for any t. Hence, always we have

$$\theta_\alpha(d) = 0. \tag{3.14}$$

We now write equations (3.5) for two symbols of differentiation d and δ:

$$d\theta_\alpha(\delta) - \delta\theta_\alpha(d) = \tilde{\omega}_\alpha^\beta(d)\theta_\beta(\delta) - \tilde{\omega}_\alpha^\beta(\delta)\theta_\beta(d). \tag{3.15}$$

By (3.14), the term with $\theta_\alpha(d)$ in (3.15) vanishes, and we obtain

$$d\theta_\alpha(\delta) = \tilde{\omega}_\alpha^\beta(d)\theta_\beta(\delta). \tag{3.16}$$

Introduce the notations

$$\theta_\alpha(\delta) = H_\alpha, \quad \tilde{\omega}_\alpha^\beta(d) = p_\alpha^\beta \, dt.$$

System (3.14) takes the form of Cauchy's system for one independent variable:

$$\frac{\partial H_\alpha}{\partial t} = p_\alpha^1 H_1 + p_\alpha^2 H_2 + \dots + p_\alpha^r H_r. \tag{3.17}$$

From equations (3.12) we find the following initial values for H_α:

$$H_\alpha = 0 \text{ for } t = 0. \tag{3.18}$$

Cauchy's system (3.17) with the initial condition (3.18) has a unique solution. This solution is obvious: since system (3.17) is homogeneous, it admits the trivial (zero) solution which satisfies condition (3.18).

Thus, we have

$$\theta_\alpha(\delta) = 0 \text{ also for } t = 1.$$

Now equations (3.13) give a homogeneous system of differential equations for the derivatives $\dfrac{\partial z^1}{\partial \alpha}, \dfrac{\partial z^2}{\partial \alpha}, \dots, \dfrac{\partial z^r}{\partial \alpha}$ at the point $t = 1$, and the determinant of this system is not equal to 0,

$$\det(c_{\alpha\beta}) \neq 0.$$

Thus, all the derivatives $\dfrac{\partial z^\beta}{\partial \alpha}$ vanish,

$$\frac{\partial z^\beta}{\partial \alpha} = 0,$$

and the values of z^β at the point M_1 are path independent.   □

18. Reduction of the problem of integration of a completely integrable system to the integration of a Cauchy system.

Theorem 5. *The problem of integration of a completely integrable system can be reduced to the integration of a system of ordinary differential equations of Cauchy's type.*

Proof. Let $A(x_0^i, z_0^\alpha)$ be a generic point of the space satisfying the conditions formulated above. An integral manifold passing through the point A admits analytic representation (3.3). In the n-dimensional space of parameters (x^i), we introduce a polar coordinate system with the origin at the point A by means of the equations

$$x^i - (x_0^i) = ta^i, \tag{3.19}$$

where t is a new positive coordinate. The parameters a^i take arbitrary values satisfying the condition

$$(a^1)^2 + (a^2)^2 + \ldots + (a^n)^2 = 1. \tag{3.20}$$

For every system of values a^i, when t changes, the point (x^i) describes a half-line emanating from the point A. All such half-lines fill out an n-dimensional space. These half-lines compose a continuous family (3.11) of paths. When we substitute $a^i dt$ for dx^i and dz^α for $dx^{n+\alpha}$, we obtain a system of ordinary differential equations

$$\frac{dz^\alpha}{dt} = Z^\alpha(a^i, t, z^\beta) \tag{3.21}$$

with initial conditions

$$z^\alpha = (z^\alpha)_0 \ \text{ for } \ t = 0.$$

In its solution we can set $t = 1$ and consider a^i as arbitrary quantities that are not connected by condition (3.20). Substituting values (3.19) for a^i, we obtain the equation of an integral manifold V.

   The values of unknown functions z^α obtained at every point (x^i) of a simply connected domain compose a solution of system (3.1).

   In fact, we can join a point M of this domain with points M' in a neighborhood of M by paths in such a way that the set of tangents to these paths at M coincides with the entire bundle of straight lines with the vertex M in the space (x^i). Since the values (z^α) were obtained by integration of system (3.1), the values z^α and dz^α, taken along the above paths, satisfy system (3.1). In particular, at the point M, the values $(z^\alpha), (dz^\alpha), (x^i), (dx^i)$ correspond to the point M and any tangent from this point. As a result, we can consider dx^i as fully arbitrary variables (the differentials of independent variables), and dz^α as the total differentials of the functions (z^α). They satisfy system (3.1). Thus (z^α) compose a solution of completely integrable system (3.1) corresponding to the initial values

$$z^\alpha = (z^\alpha)_0 \quad \text{for} \quad x^i = (x^i)_0.$$

19. First integrals of a completely integrable system. Consider a completely integrable system

$$\theta_\alpha = 0, \quad \alpha = 1, \dots, r. \tag{3.22}$$

The general solution of such a system

$$z^\alpha = f^\alpha(x^1, \dots, x^n; u^1, \dots, u^r) \tag{3.23}$$

contains r arbitrary constants—initial values of the unknown functions

$$u^\alpha = z_0^\alpha \quad \text{for} \quad x^i = (x^i)_0, \quad i = 1, \dots, n. \tag{3.24}$$

System (3.23) contains two series of values of the unknown functions z^α at the point M_0. They are of equal status. Thus system (3.23) can be solved with respect to the constants u^α,

$$u^\alpha = F^\alpha(x^1, \dots, x^n; z^1, \dots, z^r). \tag{3.25}$$

The functions u^α become constants when one replaces z^α by their values in the form of functions of all x^i, $i = 1, \dots, n$. These functions u^α are called *first integrals* of the system. System (3.22) itself is equivalent to the system

$$dF^\alpha = 0, \quad \alpha = 1, \dots, r. \tag{3.26}$$

Hence system (3.22) is completely integrable if it is algebraically equivalent to a system of form (3.26).

20. Relation between exterior differentials and the Stokes formula. As is well-known from calculus in three-dimensional space, the Stokes formula may be written as

$$\oint_L (P dx + Q dy + R dz) = \iint_S \left[\left(\frac{\partial R}{\partial y} - \frac{\partial Q}{\partial z} \right) dy dz \right.$$
$$\left. + \left(\frac{\partial P}{\partial z} - \frac{\partial R}{\partial x} \right) dz dx + \left(\frac{\partial Q}{\partial x} - \frac{\partial P}{\partial y} \right) dx dy \right], \tag{3.27}$$

where, on the right-hand side, the integral is taken over the area S of a simple surface bounded by a contour L, and on the left-hand side, the integral is taken over this contour. In the notation of an integral over a surface, for example, in

$$\iint_S \left(\frac{\partial R}{\partial y} - \frac{\partial Q}{\partial z} \right) dy dz,$$

the product of the differentials, $dy\,dz$, must be understood as an exterior product. The reason for this is that under a change of variables, for example, under the change

$$y = g(u, v) \quad z = h(u, v),$$

the product $dy\,dz$ is transformed according to the formula

$$dy\,dz \to \frac{\partial(y, z)}{\partial(u, v)}\,du\,dv = \begin{vmatrix} \dfrac{\partial g}{\partial u} & \dfrac{\partial h}{\partial u} \\[2ex] \dfrac{\partial h}{\partial v} & \dfrac{\partial h}{\partial v} \end{vmatrix}\,du\,dv,$$

and this precisely corresponds to the law of transformation of the exterior product:

$$dy \wedge dz = \left(\frac{\partial g}{\partial u}du + \frac{\partial g}{\partial v}dv\right) \wedge \left(\frac{\partial h}{\partial u}du + \frac{\partial h}{\partial v}dv\right)$$

$$= \left(\frac{\partial g}{\partial u}\frac{\partial h}{\partial v} - \frac{\partial g}{\partial v}\frac{\partial h}{\partial u}\right)du \wedge dv.$$

In this form we notice a remarkable fact. Set

$$\omega = Pdx + Qdy + Rdz$$

and compute the exterior differential of this form:

$$d\omega = \left(\frac{\partial P}{\partial y}dy + \frac{\partial P}{\partial z}dz\right) \wedge dx + \left(\frac{\partial Q}{\partial x}dx + \frac{\partial Q}{\partial z}dz\right) \wedge dy$$

$$+ \left(\frac{\partial R}{\partial x}dx + \frac{\partial R}{\partial y}dy\right) \wedge dz.$$

Collect the terms with the products $dy \wedge dz$, $dz \wedge dx$, and $dx \wedge dy$. This gives

$$d\omega = \left(\frac{\partial R}{\partial y} - \frac{\partial Q}{\partial z}\right)dy \wedge dz + \left(\frac{\partial P}{\partial z} - \frac{\partial R}{\partial x}\right)dz \wedge dx$$

$$+ \left(\frac{\partial Q}{\partial x} - \frac{\partial P}{\partial y}\right)dx \wedge dy.$$

This expression is precisely the integrand on the right-hand side of equation (3.27).

As a result, Stokes' formula (3.27) can be written in the form

$$\oint_L \omega = \iint_S d\omega. \tag{3.28}$$

One can also see the validity of this formula from the following infinitesimal considerations.

Consider a closed contour bounding a domain of a surface

$$x = f(u, v), \quad y = g(u, v), \quad z = h(u, v).$$

We decompose the entire region of integration into curvilinear parallelograms defined by intersections of coordinate lines. In order to compute a line integral along the contour, we can add the integrals along the contours of each elementary parallelogram. When we make these additions, the integrals along internal boundary lines that we pass twice in the opposite directions, cancel in pairs, and the integrals along the areas are added. Thus, it is sufficient to prove formula (3.28) for an elementary parallelogram.

21. Orientation. The Stokes formula (3.28) assumes a consistent choice of a positive passage along a contour for a line integral and of a positive side of a surface for a surface integral.

At a point M of a surface, we can establish a positive direction of rotation. This positive direction can be incrementally continuously transferred from point to point.

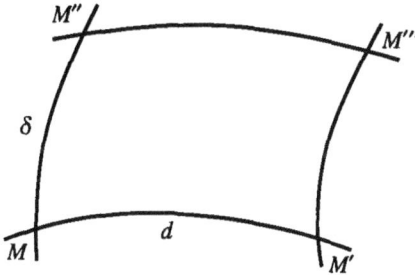

Fig. 1

Denote by M, M', M'', and M''' the vertices of a curvilinear parallelogram (see Fig. 1) such that the displacement MM' along a line u and the displacement MM'' along a line v correspond to the symbols of differentiation

$$d = \alpha \frac{\partial}{\partial u}, \quad \delta = \beta \frac{\partial}{\partial v},$$

$$\int_M^{M'} = \omega(d), \quad \int_M^{M''} = -\omega(\delta).$$

The displacement $M'M'''$ is also along a line v, and $M''M'''$ is along a line u but with new values of u and v. So, we can set

$$\int_{M'}^{M'''} = \omega(\delta) + d\omega(\delta), \quad \int_{M'''}^{M''} = -\omega(d) - \delta\omega(d).$$

Adding the four integrals along $MM', M'M''', M'''M'', M''M$, we obtain the value of the line integral along the parallelogram $MM'M'''M''M$:

$$\int_{MM'M'''M''M} \omega = \omega(d) + [\omega(\delta) + d\omega(\delta)] + [-\omega(d) - \delta\omega(d)] - \omega(\delta)$$

$$= d\omega(\delta) - \delta\omega(d),$$

and this coincides with an element of the surface integral on the right-hand side of equation (3.28). If this direction of rotation is preserved in our passage along each closed contour, then a surface is called *oriented* (two-sided).

One can also make a surface oriented by associating a bivector (a set of two tangents taken in a definite order) to any of its points.

After a surface is oriented, one can make a consistent orientation of a contour (to choose a positive orientation on a contour). Take a point M on a contour and direct the first vector emanating from M outward from the domain bounded by the contour, and direct the second one along a tangent in the positive direction of the contour. Then the orientation of the contour and the surface are consistent if the bivector constructed is positive with respect to the bivectors chosen for the surface orientation.

Chapter 4

Generalization

22. Exterior differential forms of arbitrary order. Consider an exterior quadratic form

$$\omega = a_{ij} dx^i \wedge dx^j$$

and the associated bilinear form

$$\omega(d, \delta) = \sum_{(i,j)} a_{ij} (dx^i \, \delta x^j - dx^j \, \delta x^i).$$

The expression

$$d_1 \omega(d_2, d_3) + d_2 \omega(d_3, d_1) + d_3 \omega(d_1, d_2), \tag{4.1}$$

written for three symbols of differentiation d_1, d_2, d_3, does not contain second-order differentials. In order to prove this, it is sufficient to consider just one term

$$\omega = A \, dx \wedge dy = A \begin{vmatrix} dx & dy \\ \delta x & \delta y \end{vmatrix}.$$

Expression (4.1) contains only the following terms with second-order differentials:

$$A \begin{vmatrix} d_1 d_2 x & d_1 d_2 y \\ d_3 x & d_3 y \end{vmatrix} + A \begin{vmatrix} d_2 d_3 x & d_2 d_3 y \\ d_1 x & d_1 y \end{vmatrix} + A \begin{vmatrix} d_3 d_1 x & d_3 d_1 y \\ d_2 x & d_2 y \end{vmatrix}$$

$$+ A \begin{vmatrix} d_2 x & d_2 y \\ d_1 d_3 x & d_1 d_3 y \end{vmatrix} + A \begin{vmatrix} d_3 x & d_3 y \\ d_2 d_1 x & d_2 d_1 y \end{vmatrix} + A \begin{vmatrix} d_1 x & d_1 y \\ d_3 d_2 x & d_3 d_2 y \end{vmatrix}.$$

By permutability of symbols of differentiation, we have

$$d_1 \, d_2 x \, d_3 y - d_2 \, d_1 x \, d_3 y = 0.$$

In a similar way, other pairs of terms cancel out, and the remaining terms can be written as

$$\begin{vmatrix} d_1 A & d_2 A & d_3 A \\ d_1 x & d_2 x & d_3 x \\ d_1 x & d_2 x & d_3 x \end{vmatrix} = dA \wedge dx \wedge dy.$$

This is the *trilinear covariant*, or the *exterior differential of the quadratic form* ω:

$$d\omega = \sum_{(i,j)} da_{ij} \wedge dx \wedge dy. \tag{4.2}$$

It follows that the *rule of exterior differentiation* of an exterior form of any order is: if an exterior form of order p has the general term

$$A\, dx^1 \wedge dx^2 \wedge \ldots \wedge dx^p,$$

then the exterior differential $d\omega$ of order $p+1$ has the general term

$$dA \wedge dx^1 \wedge dx^2 \wedge \ldots \wedge dx^p.$$

Exterior multiplication of exterior forms of arbitrary orders. Under exterior multiplication of two exterior forms of orders p and q, each term of one form is multiplied by each term of another with preservation of order of factors, and under exterior multiplication of monomials, their coefficients are multiplied and linear forms of each monomial are written with preservation of their orders.

If we take the exterior product of exterior forms of orders p and q,

$$\omega \text{ with the general term } A dx^1 \wedge dx^2 \wedge \ldots \wedge dx^p$$

and

$$\bar{\omega} \text{ with the general term } B dy^1 \wedge dy^2 \wedge \ldots \wedge dy^q,$$

then we obtain an exterior form of order $p+q$,

$$\omega \wedge \bar{\omega} \text{ with the general term } AB dx^1 \wedge dx^2 \wedge \ldots \wedge dx^p \wedge dy^1 \wedge dy^2 \wedge \ldots \wedge dy^q. \tag{4.3}$$

Rule of differentiation of the product of exterior forms. Suppose that we differentiate product (4.3). The general term of the exterior differential is

$$d\,(AB) \wedge dx^1 \wedge dx^2 \wedge \ldots \wedge dx^p \wedge dy^1 \wedge dy^2 \wedge \ldots \wedge dy^q$$
$$= B(dA \wedge dx^1 \wedge dx^2 \wedge \ldots \wedge dx^p \wedge dy^1 \wedge dy^2 \wedge \ldots \wedge dy^q)$$
$$+ A(dB \wedge dx^1 \wedge dx^2 \wedge \ldots \wedge dx^p \wedge dy^1 \wedge dy^2 \wedge \ldots \wedge dy^q).$$

The first term can be written in the form

$$(dA \wedge dx^1 \wedge dx^2 \wedge \ldots \wedge dx^p) \wedge (B dy^1 \wedge dy^2 \wedge \ldots \wedge dy^q) = d\omega \wedge \bar{\omega}.$$

If, in the second term, we consecutively interchange dB with the differentials $dx^1, \ldots, dx^p$, changing the sign with every interchange, then this second term is transformed into

$$A(dB \wedge dx^1 \wedge dx^2 \wedge \ldots \wedge dx^p \wedge dy^1 \wedge dy^2 \wedge \ldots \wedge dy^q)$$
$$= (-1)^p (A \wedge dx^1 \wedge dx^2 \wedge \ldots \wedge dx^p) \wedge (dB \wedge dy^1 \wedge dy^2 \wedge \ldots \wedge dy^q)$$
$$= (-1)^p \omega \wedge d\overline{\omega}.$$

Thus,

$$d\omega \wedge \overline{\omega} = d\omega \wedge \overline{\omega} + (-1)^p \omega \wedge d\overline{\omega}, \qquad (4.4)$$

where p is the order of the first form ω.

23. The Poincaré Theorem.

Theorem 1 (Poincarés). *The second exterior differential of a differential form is identically zero.*

Proof. Consider an arbitrary term of a form ω:

$$A dx^1 \wedge dx^2 \wedge \ldots \wedge dx^p.$$

Its exterior differential is

$$dA \wedge dx^1 \wedge dx^2 \wedge \ldots \wedge dx^p. \qquad (4.5)$$

If A is a function of variables $x^1, x^2, \ldots, x^p$, then dA is a linear combination of the differentials $dx^1, dx^2, \ldots, dx^p$, and exterior product (4.5) vanishes. If A is functionally independent of $x^1, x^2, \ldots, x^p$, i.e., if dA contains at least one independent variable x^{p+1}, then we can change a basis and consider dA as the differential of the new independent variable.

Since the coefficient in the product of differentials is now equal to 1 and in computing the exterior differential we multiply the exterior product from the left by the differential of this coefficient, and the latter differential is 0, we obtain

$$d(d\omega) = 0. \qquad (4.6)$$

24. The Gauss formula. The Gauss formula states:

$$\iint_F (A \, dy \, dz + B \, dz \, dx + C \, dx \, dy) = \iiint_D \left(\frac{\partial A}{\partial x} + \frac{\partial B}{\partial y} + \frac{\partial C}{\partial z} \right) dx \, dy \, dz, \quad (4.7)$$

where D is an oriented region in a three-dimensional space, and F is the consistently oriented boundary of D.

The question is how to extend this formula to an n-dimensional space. In a space of n dimensions, we consider a three-dimensional region D (i.e., we assume that there exists a system of parameters u, v, w such that to any point from D, one can put to 1-to-1 correspondence a triple of numbers (u, v, w)) bounded by a two-dimensional surface F. If ω is an exterior quadratic form, then we have

$$\iint_F \omega = \iiint_D d\omega. \tag{4.8}$$

Suppose that the region D is orientable and homeomorphic to an analytic cube, i.e., one can choose parameters u, v, w in such a way that they establish a 1-to-1 continuous correspondence between points of the region D and points of an analytic cube. An orientation can be provided by a trivector formed by a coordinate trihedron, directions of three axes of which correspond to the positive increments du, dv, dw (see Fig. 2).

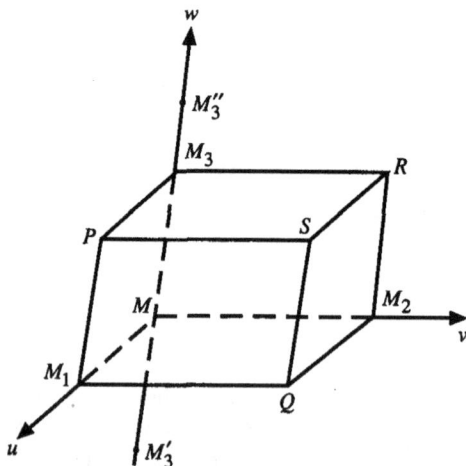

Fig. 2

Define three symbols of differentiation:

$$d_1 = \frac{\partial}{\partial u}, \quad d_2 = \frac{\partial}{\partial v}, \quad d_3 = \frac{\partial}{\partial w}.$$

The exterior differential $d\omega$ of a trilinear form can be written as

$$d\omega(d_1, d_2, d_3) = d_1\omega(d_2, d_3) + d_2\omega(d_3, d_1) + d_3\omega(d_1, d_2).$$

After integration over the volume of the parallelepiped $MM_1M_2M_3$, we obtain

$$\iiint_{MM_1M_2M_3} d\omega(d_1, d_2, d_3) = - \iiint_{MM_2RM_3} \omega(d_2, d_3) + \iiint_{M_1QSP} \omega(d_2, d_3)$$

$$- \iiint_{MM_1PM_3} \omega(d_3, d_1) + \iiint_{M_1QSR} \omega(d_3, d_1) - \iiint_{MM_1QM_2} \omega(d_1, d_2) + \iiint_{M_1PSR} \omega(d_1, d_3).$$

The coordination of signs: if the bivector MM_1M_2 is negative, then the trivector $MM_3'M_1M_2$ is negative; if the bivector M_3PQ is positive, then the trivector $MM_3''PQ$ is also positive, etc.

This transformation can be extended to any dimension. Let D be a region of dimension $p+1$ and F be its boundary of p dimensions. Suppose that D admits an orientation, i.e., one can define a positive numbering of frame vectors at any point which can be continuously transferred from point to point. Then an orientation of the region D can be transferred to its boundary F: namely, if a point M belongs to the boundary F, $\overrightarrow{MT}$ is an outward vector (it is directed from the point M to the external part of the space), and $\overrightarrow{MT_1}, \overrightarrow{MT_2}, \dots, \overrightarrow{MT_p}$ are vectors inside p-dimensional manifold F, then the multivector $MT_1 \dots T_p$ is positive if including multivector $MTT_1 \dots T_p$ is positive. Under these conditions, we have

$$\int\limits_{F}^{(p)} \omega = \int\limits_{D}^{(p+1)} d\omega. \tag{4.9}$$

25. Generalization of Theorem 6 of No. 12.

Theorem 2. *The exterior product of a form F and a linear form f vanishes if and only if F is a product of f and an appropriate form φ.*

Proof. The theorem condition is sufficient. If

$$F = f \wedge \varphi,$$

then

$$F \wedge f = f \wedge \varphi \wedge f = 0,$$

since each term of the associated multilinear form has the form of a determinant with two equal rows.

The theorem condition is necessary. Replace a basis of linear forms by a new basis containing the form f:

$$f, f^1, \dots, f^{n-1}.$$

If we assume that in the new basis, after collecting similar terms, there is a term not containing f,

$$a_{12\dots p}f^1 \wedge f^2 \wedge \dots \wedge f^p, \tag{4.10}$$

then the product on the left-hand side of the equation

$$F \wedge f = 0$$

contains the term

$$a_{12\dots p}f^1 \wedge f^2 \wedge \dots \wedge f^p \wedge f,$$

which is not 0. This follows from the fact that all the factors $f^1, f^2, \dots, f^p, f$ are different basis forms, and the above written term is not similar to any other term of the product $F \wedge f$ since the form F has only one term (4.10) containing the indicated basis terms. Thus all terms of F contain f.   □

EXAMPLE. Consider the form

$$\omega = P\,dx + Q\,dy + R\,dz.$$

The equation $\omega = 0$ is completely integrable if $d\omega = 0$ is an algebraic consequence of $\omega = 0$.

By Theorem 2, this condition is equivalent to the equation

$$\omega \wedge d\omega = 0.$$

Since

$$d\omega = \left(\frac{\partial R}{\partial y} - \frac{\partial Q}{\partial z}\right) dy \wedge dz + \left(\frac{\partial P}{\partial z} - \frac{\partial R}{\partial x}\right) dz \wedge dx$$
$$+ \left(\frac{\partial Q}{\partial x} - \frac{\partial Q}{\partial y}\right) dx \wedge dy,$$

$$\omega \wedge d\omega = \left\{ P\left(\frac{\partial R}{\partial y} - \frac{\partial Q}{\partial z}\right) + Q\left(\frac{\partial P}{\partial z} - \frac{\partial R}{\partial x}\right) \right.$$
$$\left. + R\left(\frac{\partial Q}{\partial x} - \frac{\partial P}{\partial y}\right) \right\} dx \wedge dy \wedge dz,$$

the condition of complete integrability is the vanishing of the expression in brackets.

A. GEOMETRY OF EUCLIDEAN SPACE

Chapter 5

The Existence Theorem for a Family of Frames with Given Infinitesimal Components ω^i and ω_i^j

26. Family of oblique trihedrons. We return to geometric problems. In the beginning of Chap. 2, we posed the problem: *given six forms ω^i and ω_i^j of an infinitesimal displacement of an orthonormal trihedron, does there exist a family of trihedrons admitting these components ?*

It is convenient to consider first the case of oblique trihedrons when the frame vectors are not connected by any additional conditions except the condition of linear independence.

Consider system (1.9):

$$\begin{cases} d\mathbf{M} = \omega^i \mathbf{e}_i, \\ d\mathbf{e}_i = \omega_i^j \mathbf{e}_j \end{cases} \tag{I}$$

and assume that forms ω^i and ω_i^j are given as linear combinations of the differentials du^α with coefficients of class C^1. If system (I) admits a solution, then substituting this solution into equations of (I), we obtain an identity which can be differentiated term by term.

We apply exterior differentiation to both parts of system (I) under this assumption. Then the left-hand side is identically equal to 0 (the exterior differential of a total differential), and as a result, for the right-hand side we have

$$d\mathbf{e}_i \wedge \omega^i + \mathbf{e}_i d\omega^i = 0,$$
$$d\mathbf{e}_j \wedge \omega_i^j + \mathbf{e}_j d\omega_i^j = 0,$$

or substituting de_i and de_j from (I), we find that

$$\omega_i^j \wedge \omega^i \mathbf{e}_j + \mathbf{e}_i d\omega^i = 0,$$
$$\omega_j^k \wedge \omega_i^j \mathbf{e}_k + \mathbf{e}_j d\omega_i^j = 0.$$

Changing the summation index in the first two terms of these equations (in the first equation i for j, and in the second one k for j) and taking the common factor ($\mathbf{e}_i$ in the first equation and $\mathbf{e}_j$ in the second one) across the brackets, we obtain

$$\begin{cases} \mathbf{e}_i(d\omega^i + \omega_j^i \wedge \omega^j) = 0, \\ \mathbf{e}_j(d\omega_i^j + \omega_k^j \wedge \omega_i^k) = 0. \end{cases} \qquad (5.1)$$

Since the basis vectors $\mathbf{e}_i$ are linearly independent, the expressions in brackets are equal to 0. Hence

$$\begin{cases} d\omega^i = \omega^j \wedge \omega_j^i, \\ d\omega_i^j = \omega_i^k \wedge \omega_k^j. \end{cases} \qquad (II)$$

Equations (I) are *the structure equations of the space*. Solutions of system (I) (if they exist) must satisfy equations (II). We prove that these conditions are not only necessary but also sufficient for system (I) to define a family of trihedrons.

Theorem 1. *If linear forms ω^i and ω_i^j satisfy structure equations (II), then system (I) is completely integrable, and for any initial position of a nondegenerate trihedron T_0 (a set of vectors $\mathbf{M}_0$ and $\mathbf{e}_i^0$ at a point M_0), they define a family of frames obtained from an initial trihedron T_0 by an appropriate affine transformation.*

Proof. In fact, we saw that exterior differentiation of system (I) (with the use of system (I) itself) leads to system (5.1). The equations of the latter system become the identities, if the forms ω^i and ω_i^j satisfy structure equations (II). Thus, under the theorem conditions, system (I) is completely integrable, and for any initial values of coordinates of the vector $\mathbf{M}_0$ and independent vectors $\mathbf{e}_i^0$ given for $u^\alpha = u_0^\alpha$, there exists a solution of (I) with the number of independent variables u^α (not exceeding 12) which is equal to the number of these parameters occurring in the given differential forms ω^i and ω_i^j. $\qquad \square$

    27. The family of orthonormal trihedrons. Equations (1.1) of an infinitesimal displacement of an orthonormal trihedron are a particular case of general system (1.9). Thus the existence theorem which we just proved can be expanded to the system of equations (1.1). However, in the case of orthonormal trihedrons, we have additional conditions (1.3). We will now prove that if an initial trihedron (T_0) satisfies conditions (1.3), then all trihedrons T of an integral manifold satisfy these conditions. We write conditions (1.3) in a more compact form

$$\mathbf{I}_i \cdot \mathbf{I}_j = \delta_{ij} = \begin{cases} 0, & \text{if } i \neq j, \\ 1, & \text{if } i = j. \end{cases} \qquad (5.2)$$

We assume that conditions (5.2) are satisfied at the point M_0, i.e., for $u^\alpha = u_0^\alpha$. Set

$$\mathbf{I}_i \cdot \mathbf{I}_j = g_{ij}. \tag{5.3}$$

Differentiating this identity and applying equations (1.1), we find that

$$dg_{ij} = \mathbf{I}_j \cdot d\mathbf{I}_i + \mathbf{I}_i \cdot d\mathbf{I}_j = \mathbf{I}_j \cdot (\omega_i^k \mathbf{I}_k) + \mathbf{I}_i \cdot (\omega_j^k \mathbf{I}_k),$$

or, by (5.3),

$$dg_{ij} = \omega_i^k g_{kj} + \omega_j^k g_{ik}. \tag{5.4}$$

For given forms ω_i^j, this is a Pfaffian system for the unknown functions g_{ij}. We saw (see Theorem 2 of No. **16**) that such a system admits no more than one solution with the initial conditions

$$g_{ij} = \delta_{ij} \ \text{ for } \ u^\alpha = u_0^\alpha. \tag{5.5}$$

However, one such solution is obvious—it is determined now by equations (5.5) for all values of variables u^α. After substituting (5.5) into (5.4), we find the left-hand side equals 0, and as a result, for the right-hand side we have

$$\omega_i^k \delta_{kj} + \omega_j^k \delta_{ik} = 0,$$

or

$$\omega_i^j + \omega_j^i = 0,$$

and from (1.4), the latter equations are satisfied. Thus, it is sufficient to choose the initial trihedron to be orthonormal in order to have all trihedrons orthonormal. This solves our problem.

28. Family of oblique trihedrons with a given line element. An orthonormal trihedron has its basis vectors $\mathbf{I}_i$ mutually orthogonal and of unit length. If such trihedrons are of the same orientation (for instance, if all of them are right trihedrons, as we agreed in No. 1), then they are congruent. Thus, we say that the system of equations (1.1), under conditions (1.4) and (II), defines a motion of a trihedron in the space, and the system of equations (1.1), under conditions (II) only, defines an infinitesimal affine transformation of a trihedron.

However, one should not think that this conclusion is necessarily connected with the frame orthonormality. It is easy to note that equations (1.4) are a particular case of the system (5.4) and can be obtained from (5.4) by means of relations (1.3). If we keep the system (5.4) and take an arbitrary (nondegenerate) initial frame (i.e., for $u^\alpha = u_0^\alpha$, we take arbitrary values of vectors $\mathbf{M}_0$ and $\mathbf{e}_i^0$, the only condition being that the vectors $\mathbf{e}_i^0$ do not lie in a plane), then to different initial frames, there exist corresponding affine transformations of the resulting family of trihedrons since there exists one and only one affine transformation of the space sending the initial trihedron to an arbitrarily assigned (nondegenerate) trihedron. Moreover, if initial frames coincide, then the two resulting families of trihedrons coincide, since a linear affine transformation of vectors with constant coefficients applied simultaneously to both parts of equations (1.9) does not change relative components ω^i and ω_i^j.

If we transform the initial frame preserving the initial values of coefficients (1.10),

$$g_{ij} = g_{ji} \text{ for } u^\alpha = u_0^\alpha,$$

then system (5.4) produces the same solution, and at every point of the space, the line element (1.12) remains the same. All the distances are preserved, and the transformation of the integral manifold of frames is simply a translation (cf. No. 32).

29. Integration of system (I) by the method of the form invariance.

Consider all orthonormal trihedrons of the ordinary Euclidean space. They are defined by three coordinates x, y, z of their vertex M and by three Euler angles θ, φ, ψ determining the rotation of the trihedron.

For these orthonormal trihedrons, we compute the components

$$\widetilde{\omega}^i(x, y, z, \varphi, \psi; dx, dy, dz, d\theta, d\varphi, d\psi),$$

$$\widetilde{\omega}_i^j(x, y, z, \theta, \varphi, \psi; dx, dy, dz, d\theta, d\varphi, d\psi)$$

of an infinitesimal displacement. These forms are linear forms corresponding to any real motion of an orthonormal trihedron in the space. Thus, they satisfy the structure equations

$$\begin{cases} d\widetilde{\omega}^i = \widetilde{\omega}^k \wedge \widetilde{\omega}_k^i, \\ d\widetilde{\omega}_i^j = \widetilde{\omega}_i^k \wedge \widetilde{\omega}_k^j. \end{cases} \tag{5.6}$$

If there exists a solution of the system of equations (1.1) for the forms ω^i and ω_i^j that corresponds to a motion of the trihedron in the space, then there must exist functions $x, y, z, \theta, \varphi, \psi$ of the variables u^α such that the equations

$$\begin{cases} \widetilde{\omega}^i(x, \ldots, \psi; dx, \ldots, d\psi) = \omega^i(u^1, \ldots, u^6; du^1, \ldots, du^6), \\ \widetilde{\omega}_i^j(x, \ldots, \psi; dx, \ldots, d\psi) = \omega_i^j(u^1, \ldots, u^6; du^1, \ldots, du^6). \end{cases} \tag{5.7}$$

are satisfied. System (5.7) is obviously a system of Pfaffian equations. Applying exterior differentiation to both parts of these equations, we find that

$$\begin{cases} d\widetilde{\omega}^i = d\omega^i, \\ d\widetilde{\omega}_i^j = d\omega_j^i, \end{cases}$$

or, by (II) and (5.6), we get

$$\begin{cases} \widetilde{\omega}^k \wedge \widetilde{\omega}_k^i = \omega^k \wedge \omega_k^i, \\ \widetilde{\omega}_i^k \wedge \widetilde{\omega}_k^j = \omega_i^k \wedge \omega_k^j. \end{cases}$$

Obviously, by (5.7), the last equations are satisfied. Thus, system (5.7) is completely integrable, and for any initial location of a trihedron,

$$x = x^0, \quad y = y^0, \quad z = z^0,$$
$$\theta = \theta^0, \quad \varphi = \varphi^0, \quad \psi = \psi^0 \text{ for } u^\alpha = u_0^\alpha.$$

there exists a unique solution defining the motion of the trihedron T.

Remark. The method of comparison of components ω and $\tilde{\omega}$ we have used above is a method for determining transformations in the group of motions of a trihedron. Equations (II) are the structure equations of the group, and the forms ω^i and ω_i^j are invariant forms of this group.

30. Particular cases. *Euclidean plane.* We have two axes $\mathbf{I}_1, \mathbf{I}_2$, three forms $\omega^1, \omega^2, \omega_1^2$, and three parameters x, y, θ (see Fig. 3).

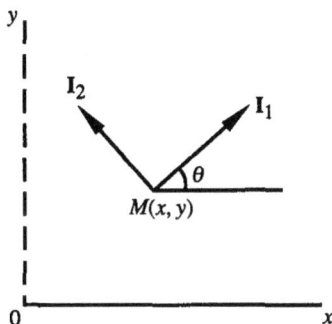

Fig. 3

The forms ω^1, ω^2 are the projections of the vector (dx, dy) onto the axes $\mathbf{I}_1, \mathbf{I}_2$. We have

$$\begin{cases} \omega^1 = dx \, \cos\theta + dy \, \sin\theta, \\ \omega^2 = -dx \, \sin\theta + dy \, \cos\theta, \\ \omega_1^2 = d\theta. \end{cases} \tag{5.8}$$

Conditions (II) are satisfied:

$$\begin{cases} d\omega^1 = \omega^2 \wedge \omega_2^1, \\ d\omega^2 = \omega^1 \wedge \omega_1^2, \\ d\omega_1^2 = 0. \end{cases}$$

In fact,

$$d\omega^1 = -\sin\theta\, d\theta \wedge dx + \cos\theta\, d\theta \wedge dy = d\theta \wedge \omega^2 = \omega_1^2 \wedge \omega^2 = \omega^2 \wedge \omega_2^1,$$

and the two other conditions can be checked in a similar way.

EXAMPLE. Define the forms

$$\begin{cases} \omega^1 = du - v\,dw, \\ \omega^2 = dv + u\,dw, \\ \omega_1^2 = dw. \end{cases} \tag{5.9}$$

Equations (II) are satisfied. Thus, these three forms define an orthonormal frame if the initial frame is orthonormal. We apply the second method. Comparing equations (5.8) with (5.9), we find that

$$\begin{cases} dx\, \cos\theta + dy\, \sin\theta = du - v\,dw, \\ -dx\, \sin\theta + dy\, \cos\theta = dv + u\,dw, \\ d\theta = dw. \end{cases}$$

Integrating the last equation, we obtain

$$\theta = w + c, \quad c = \text{const.}$$

From the first two equations, we find dx:

$$dx = \cos(w+c)du - \sin(w+c)dv$$
$$-[v\cos(w+c) + u\sin(w+c)]dw.$$

Integrating, we obtain

$$x = u\cos(w+c) - v\sin(w+c) + a, \quad a = \text{const.}$$

Similarly, we have

$$y = u\sin(w+c) + v\cos(w+c) + b, \quad b = \text{const.}$$

Equations (II) are *the structure equations of the group of motions.*

Darboux's theory of surfaces. Darboux considered motions depending on two parameters. He defined

$$\begin{cases} \omega^1 = \xi\,du + \xi_1\,dv, & \omega_2^3 = p\,du + p_1\,dv, \\ \omega^2 = \eta\,du + \eta_1\,dv, & \omega_2^3 = q\,du + q_1\,dv, \\ \omega^3 = \zeta\,du + \zeta_1\,dv, & \omega_2^3 = r\,du + r_1\,dv. \end{cases}$$

Then conditions (II) take the following form:
THE FIRST SERIES of equations (II) is

$$
\begin{cases}
d\omega^1 = \omega^2 \wedge \omega_2^1 + \omega^3 \wedge \omega_3^1, \\
\dfrac{\partial \xi_1}{\partial u} - \dfrac{\partial \xi}{\partial v} = (\eta r_1 - \eta_1 r) + (\zeta q_1 - \zeta_1 q),
\end{cases}
\tag{5.10}
$$

and the others can be obtained by a simultaneous cyclic permutation of two series of letters, ξ, η, ζ and p, q, r.
THE SECOND SERIES of equations (II) is

$$
\begin{cases}
d\omega_2^3 = \omega_2^1 \wedge \omega_3^1, \\
\dfrac{\partial p_1}{\partial u} - \dfrac{\partial p}{\partial v} = r q_1 - r_1 q,
\end{cases}
\tag{5.11}
$$

and the others can be obtained by a cyclic permutation of letters p, q, r.

31. Spaces of trihedrons. *A family of trihedrons with a common origin.* How should we define such a family by means of differential equations?
Since the origin M is fixed, we have

$$
\omega^1 = 0, \quad \omega^2 = 0, \quad \omega^3 = 0.
\tag{5.12}
$$

This is a system of three Pfaffian equations with three unknown functions of three independent variables. The system is completely integrable. This is obvious from geometric considerations (such a family does exist) but it is also possible to check directly that equations (II) are identically satisfied.
A family of trihedrons with a common plane. Such a family is defined by the system

$$
\omega^3 = 0, \quad \omega_1^3 = 0, \quad \omega_2^3 = 0.
$$

The system is completely integrable.
A family of trihedrons with a common edge $\mathbf{I}_3$. Such a family is defined by the system

$$
\omega^1 = 0, \quad \omega^2 = 0, \quad \omega_1^3 = 0, \quad \omega_2^3 = 0.
\tag{5.13}
$$

This configuration depends on four arbitrary parameters. This corresponds to the well-known fact that the three-dimensional point space is four-dimensional if it is considered as a ruled space (i.e., a space whose generating element is a straight line).
A point as a configuration of trihedrons. A point as a configuration of trihedrons can be also defined in a different way.
Consider a trihedron and a point A with the coordinates a^1, a^2, a^3. We request that the point A is fixed under trihedron's motions inside a family of trihedrons.

We have

$$\mathbf{A} = \mathbf{M} + a^1\mathbf{I}_1 + a^2\mathbf{I}_2 + a^3\mathbf{I}_3.$$

Here the length of the vector $\mathbf{A}$ is the distance from the fixed origin to the point $A(a^1, a^2, a^3)$. Differentiating the last equation and applying the fact that the point A is fixed, we find

$$d\mathbf{M} + a^1 d\mathbf{I}_1 + a^2 d\mathbf{I}_2 + a^3 d\mathbf{I}_3 = 0.$$

Substituting the values of $d\mathbf{M}$ and $d\mathbf{I}_i$ from equations (1.1) into this equation and equating the coefficients in the linearly independent vectors $\mathbf{I}_1, \mathbf{I}_2, \mathbf{I}_3$ to 0, we obtain

$$\begin{cases} \omega^1 + a^2\omega_2^1 + a^3\omega_3^1 = 0, \\ \omega^2 + a^3\omega_3^2 + a^1\omega_1^2 = 0, \\ \omega^3 + a^1\omega_1^3 + a^2\omega_2^3 = 0. \end{cases} \tag{5.14}$$

This system is completely integrable.

Thus, we can consider trihedrons as constructive elements of the space and define points, planes, and straight lines as configurations of trihedrons.

How can we prove that different definitions of a point are equivalent?

Set

$$\begin{cases} \widetilde{\omega}^1 = \omega^1 + a^2\omega_2^1 + a^3\omega_3^1, \\ \widetilde{\omega}^2 = \omega^2 + a^3\omega_3^2 + a^1\omega_1^2, \\ \widetilde{\omega}^3 = \omega^3 + a^1\omega_1^3 + a^2\omega_2^3, \\ \widetilde{\omega}_2^3 = \omega_2^3, \\ \widetilde{\omega}_3^1 = \omega_3^1, \\ \widetilde{\omega}_1^2 = \omega_1^2. \end{cases}$$

The new forms $\widetilde{\omega}^i$ and $\widetilde{\omega}_i^j$ satisfy conditions (II), and the point is defined by the system of equations

$$\widetilde{\omega}^1 = 0, \quad \widetilde{\omega}^2 = 0, \quad \widetilde{\omega}^3 = 0. \tag{5.15}$$

Consider two systems of parameters $x, y, z, \theta, \varphi, \psi$ and $x', y', z', \theta', \varphi', \psi'$. Set $\widetilde{\omega} = \omega(x', y', z', \theta', \varphi', \psi')$. The system

$$\widetilde{\omega}^i = \omega^i, \quad \widetilde{\omega}_i^j = \omega_i^j \tag{5.16}$$

is completely integrable. System (5.16) is a system of 6 equations for 6 unknown functions $x', y', z', \theta', \varphi', \psi'$ with 6 independent variables $x, y, z, \theta, \varphi, \psi$,

Assigning initial values, we find that to any motion of the trihedron $T(x, y, z, \theta, \varphi, \psi)$ into a trihedron $T_1(x_1, y_1, z_1, \theta_1, \varphi_1, \psi_1)$, there is a corresponding motion of the trihedron $T'(x', y', z', \theta', \varphi', \psi')$ into a trihedron $T_1'(x_1', y_1', z_1', \theta_1', \varphi_1', \psi_1')$, and all the forms ω^i, ω_i^j are equal to $\widetilde{\omega}^i, \widetilde{\omega}_i^j$, respectively, and define the same motion in the space.

Equations (5.16) define a set of motions in the space.

Chapter 6

The Fundamental Theorem of Metric Geometry

32. The rigidity of the point space.

Theorem 1 (Weyl). *The forms ω^1, ω^2, and ω^3, given as functions of the parameters u^α, in such a way that the vertices of trihedrons fill out a three-dimensional domain of the space (i.e., the forms ω^i are linearly independent) define the forms ω_2^3, ω_3^1, and ω_1^2 uniquely.*

Proof. We consider the general case. We prove that equations (II) of No. **26**

$$d\omega^i = \omega^j \wedge \omega_j^i \tag{II$_1$}$$

have a unique solution.

Assume that there are two solutions, ω_i^j and $\widetilde{\omega}_i^j$. Set

$$\overline{\omega}_i^j = \omega_i^j - \widetilde{\omega}_i^j.$$

Then composing the differences between the left-hand sides and the right-hand sides of two equations of (II$_1$), one for the solution ω_i^j and the second for the solution $\widetilde{\omega}_i^j$, we obtain that

$$\omega^k \wedge \overline{\omega}_i^j = 0.$$

Applying Cartan's lemma to the last exterior quadratic equation, we find that

$$\overline{\omega}_k^i = c_{kh}^i \omega^h, \tag{6.1}$$

$$c_{kh}^i = c_{hk}^i. \tag{6.2}$$

On the other hand, since the forms ω_i^j are skew-symmetric, we have

$$c_{kh}^i = -c_{ih}^k. \tag{6.3}$$

45

Now applying transformations (6.2) and (6.3) in turn, we find that

$$c_{kh}^i = c_{hk}^i = -c_{ik}^h = -c_{ki}^h = c_{hi}^k = c_{ih}^k = -c_{kh}^i.$$

It follows that

$$c_{kh}^i = 0$$

and

$$\varpi_k^i = 0, \quad \text{i.e.,} \quad \omega_k^i = \tilde{\omega}_k^i.$$

33. Geometric meaning of the Weyl theorem. We will consider a transformation of a family of trihedrons establishing a point correspondence of the space.

Consider two neighborhoods: a neighborhood of a point M of the first family and a neighborhood of the corresponding point M' of the second family. Then moving the entire first family in such a way that the point M coincides with the point M', we can make the neighborhoods of these two points coincide. Such a transformation is called a *deformation* of the space.

The Weyl theorem asserts that the only deformations of our space are translations. In other words, *the point space is rigid.*

We present some details of a proof of this statement.

To a point M_0, there corresponds a point $\widetilde{M}_0$ such that an infinitesimal part of the space surrounding the point M_0 (with the point M_0 itself) is equal to the corresponding part surrounding the point $\widetilde{M}_0$. To the point M_0, we attach a trihedron T_0 with its origin at M_0. To T_0, there corresponds a trihedron $\widetilde{T}_0$ with its vertex at $\widetilde{M}_0$.

If (ω^i) is an infinitesimal vector $\overrightarrow{M_0M}$, and $(\tilde{\omega}^i)$ is the corresponding vector $\overrightarrow{\widetilde{M}_0\widetilde{M}}$, then the corresponding transformation is defined by the equations

$$\tilde{\omega}^1 = \omega^1, \quad \tilde{\omega}^2 = \omega^2, \quad \tilde{\omega}^3 = \omega^3. \tag{6.4}$$

In fact, we can move the second family of trihedrons in such a way that the trihedron $\widetilde{T}_0$ coincides with T_0. Suppose that under this transformation, the point $\widetilde{M}$ is moved to the point M'. From equations (6.4), we see that the point M' coincides with M up to infinitesimals of second order. (It is assumed that the point M is infinitesimally close to the point M_0.)

Conversely, consider a point correspondence sending a part of the space surrounding the point M_0 to an equal part of the space surrounding the point $\widetilde{M}_0$.

Let $\widetilde{T}_0$ be a trihedron that coincides with the trihedron T_0 under a motion making two corresponding neighborhoods coincide. Suppose that the trihedron $\widetilde{T}_0$ corresponds to the trihedron T_0.

Then we have

$$\omega^i = \tilde{\omega}^i.$$

This implies that

$$\omega_i^j = \tilde{\omega}_i^j,$$

and the transformation is just a translation of the space as a rigid body.

REMARK. A property we just proved is not so obvious: this can be seen by considering the tangential space (the space of planes).

34. Deformation of the tangential space. Consider two points M and M' and trihedrons T and T' attached to these points. Suppose that a point P has coordinates x, y, z with respect to the trihedron T and coordinates x', y', z' with respect to the trihedron T'.

Then, projecting the position vector $\overrightarrow{M'M}$ onto the axes of the first trihedron T, we obtain

$$x = w^1 + x'\mathbf{I}'_1 \cdot \mathbf{I}_1 + y'\mathbf{I}'_2 \cdot \mathbf{I}_2 + z'\mathbf{I}'_3 \cdot \mathbf{I}_3 \qquad (6.5)$$

and two similar expressions for y and z.

In addition, if we assume that the points M and M' are infinitesimally close, then by equations (1.1), we have

$$\begin{cases} \mathbf{I}'_1 = \mathbf{I}_1 + d\mathbf{I}_1 = \mathbf{I}_1 + w_1^2\mathbf{I}_2 + w_1^3\mathbf{I}_3, \\ \mathbf{I}'_2 = \mathbf{I}_2 + d\mathbf{I}_2 = \mathbf{I}_2 + w_2^1\mathbf{I}_1 + w_2^3\mathbf{I}_3, \\ \mathbf{I}'_3 = \mathbf{I}_3 + d\mathbf{I}_3 = \mathbf{I}_3 + w_3^1\mathbf{I}_1 + w_3^2\mathbf{I}_2. \end{cases}$$

Thus, equations (6.5) become

$$\begin{cases} x = w^1 + x' + y'w_2^1 + z'w_3^1, \\ y = w^2 + y' + z'w_3^2 + x'w_1^2, \\ z = w^3 + z' + x'w_1^3 + y'w_2^3, \end{cases}$$

and conversely,

$$\begin{cases} x' = -w^1 + x - yw_2^1 - z'w_3^1, \\ y' = -w^2 + y - zw_3^2 - xw_1^2, \\ z' = -w^3 + z - xw_1^3 - yw_2^3. \end{cases}$$

It follows that the trihedrons T and T' have the same plane xy. The equations $z = 0$ and $z' = 0$ coincide, if

$$w^3 = 0, \quad w_1^3 = 0, \quad w_2^3 = 0.$$

Otherwise, in the general case, the equation of the plane $z' = 0$ with respect to the trihedron T is

$$z = w^3 + xw_1^3 + yw_2^3.$$

Thus, the forms

$$w^3, \quad w_1^3, \quad w_2^3$$

are the tangential coordinates of an infinitesimally close plane. These forms are linearly independent if a family of planes contains all the planes in a domain of the three-dimensional space—we will assume that this is the case.

Consider now two manifolds of planes (or two copies of one manifold of planes), and in each of these spaces consider a family of trihedrons. Suppose that an infinitesimal displacement of trihedrons of the first family is determined by the forms ω^i, ω^j_i, and that of the second family is determined by the forms $\widetilde{\omega}^i, \widetilde{\omega}^j_i$. We can also assume that the forms ω^i, ω^j_i are written for variables u^α, and the forms $\widetilde{\omega}^i, \widetilde{\omega}^j_i$ for variables v^α.

Consider a correspondence established between trihedrons of these two families by the system

$$\widetilde{\omega}^3 = \omega^3, \quad \widetilde{\omega}^3_1 = \omega^3_1, \quad \widetilde{\omega}^3_2 = \omega^3_2. \tag{6.6}$$

It is not difficult to check that this system is completely integrable. However, this system establishes a correspondence not between trihedrons of the above two spaces but between trihedrons of the two families chosen above. Namely, if in the first space, the trihedrons describe a family with the common plane xy, then

$$\omega^3 = \omega^3_1 = \omega^3_2 = 0.$$

By (6.6), this implies that

$$\widetilde{\omega}^3 = \widetilde{\omega}^3_1 = \widetilde{\omega}^3_2 = 0.$$

Thus, the corresponding trihedrons of the second space also compose a family with the common plane $x'y'$. Hence, system (6.6) establishes a correspondence between planes of the first and second spaces.

Consider now two corresponding planes P_0 and $\widetilde{P}_0$ of the first and the second spaces and two trihedrons T_0 and $\widetilde{T}_0$ corresponding to these points.

Move the second space with its planes and trihedrons in such a way that its plane $\overline{P}_0$ coincides with the plane P_0, and its trihedron $\widetilde{T}_0$ coincides with the trihedron T_0.

Consider a plane P in an infinitesimal neighborhood of the plane P_0 of the first space and the plane $\widetilde{P}$ in an infinitesimal neighborhood of the plane $\widetilde{P}_0$ of the second space corresponding to P under correspondence (6.6).

First, we note that the corresponding plane $\overline{P}$ is really infinitesimally close to the plane $\widetilde{P}_0$. The point is that with respect to the trihedron T_0, the plane P has the coordinates

$$\omega^3, \ \omega^3_1, \ \omega^3_2,$$

and by means of correspondence (6.6), the plane $\widetilde{P}$ has the coordinates

$$\widetilde{\omega}^3, \ \widetilde{\omega}^3_1, \ \widetilde{\omega}^3_2$$

with respect to the trihedron $\widetilde{T}_0$, and the latter coordinates are precisely the coordinates of the plane P. Thus, after moving the second space and having the plane $\widetilde{P}_0$ coincide with the plane P_0 and the trihedron $\widetilde{T}_0$ coincide with the trihedron T_0, the planes P and $\widetilde{P}$ (up to infinitesimals of second order) have the same coordinates with respect to the same frame T_0. Thus, these planes coincide (up to infinitesimals of second order).

In this way, the first-order deformation is described in the manifold of planes. Conversely, if two planes coincide (up to infinitesimals of second order), and the same is true for infinitesimally close planes, then the coordinates of these planes with respect to a common frame coincide (up to infinitesimals of second order). Thus, we have

$$\widetilde{\omega}^3 = \omega^3, \quad \widetilde{\omega}_1^3 = \omega_1^3, \quad \widetilde{\omega}_2^3 = \omega_2^3.$$

Does it follow from these equations that other components of infinitesimal displacement coincide and the spaces are congruent?

Consider in succession exterior differentials of the forms on both sides of equations (6.6) and apply equations (6.6) to replace the forms $\widetilde{\omega}$ by the forms ω. We have

For the trihedron T:

$$dw^3 = \omega^1 \wedge \omega_1^3 + \omega^2 \wedge \omega_2^3,$$

$$d\omega_1^3 = \omega_1^2 \wedge \omega_3^2,$$

$$d\omega_2^3 = \omega_2^1 \wedge \omega_3^1,$$

For the trihedron $\widetilde{T}$:

$$dw^3 = \widetilde{\omega}^1 \wedge \omega_1^3 + \widetilde{\omega}^2 \wedge \omega_2^3,$$

$$d\omega_1^3 = \widetilde{\omega}_1^2 \wedge \omega_3^2,$$

$$d\omega_2^3 = \widetilde{\omega}_2^1 \wedge \omega_3^1.$$

Comparing the equations of the last two rows, we obtain

$$(\widetilde{\omega}_1^2 - \omega_1^2) \wedge \omega_2^3 = 0,$$

$$(\widetilde{\omega}_2^1 - \omega_2^1) \wedge \omega_1^3 = 0.$$

By Cartan's Lemma (see No. **12**), it follows that

$$\widetilde{\omega}_1^2 - \omega_1^2 = \alpha\omega_2^3,$$

$$\widetilde{\omega}_2^1 - \omega_2^1 = \beta\omega_1^3.$$

Adding these equations termwise and applying the equation $\omega_1^2 + \omega_2^1 = 0$, we find that

$$\alpha\omega_2^3 + \beta\omega_2^3 = 0.$$

Since the forms ω_1^3 and ω_2^3 are linearly independent, from the last equation it follows that

$$\alpha = \beta = 0,$$

and as a result,

$$\widetilde{\omega}_1^2 = \omega_1^2. \tag{6.7}$$

Comparing the first row equations gives

$$(\widetilde{\omega}^1 - \omega^1) \wedge \omega_1^3 + (\widetilde{\omega}^2 - \omega^2) \wedge \omega_2^3 = 0.$$

From (2.11) and (2.12), it follows that

$$\begin{cases} \widetilde{\omega}^1 - \omega^1 = a\omega_1^3 + b\omega_2^3, \\ \widetilde{\omega}^2 - \omega^2 = b\omega_1^3 + c\omega_2^3. \end{cases} \tag{6.8}$$

Exterior differentiation of equation (6.7) leads to an identity, and exterior differentiation of equations (6.8) gives exterior quadratic equations for da, db, and dc that in general are not identities.

Thus, the deformation of the manifold of planes does not reduce to a translation. Hence, *the tangential space is less rigid than the point space*—in the former it is possible to construct a transformation not changing small pieces of the space but deeply changing the entire space.

35. Deformation of the plane considered as a locus of straight lines. Consider two planes with coordinates (x, y) and (x', y'). Applying formulas of No. **30**, in each of these planes, we can choose a rectangular coordinate system defined respectively by the bivectors $(M, \mathbf{I}_1, \mathbf{I}_2)$ and $(M', \mathbf{I}_1', \mathbf{I}_2')$. Then, in the first plane we have

$$
\begin{cases}
d\mathbf{M} = \omega^1 \mathbf{I}_1 + \omega^2 \mathbf{I}_2, \\
d\mathbf{I}_1 = \omega_1^2 \mathbf{I}_2, \\
d\mathbf{I}_2 = \omega_2^1 \mathbf{I}_1,
\end{cases}
\tag{6.9}
$$

$$
\begin{cases}
\omega^1 = dx \cos \theta + dy \sin \theta, \\
\omega^2 = -dx \sin \theta + dy \cos \theta, \\
\omega_1^2 = d\theta,
\end{cases}
\tag{6.10}
$$

and in the second plane the forms $\widetilde{\omega}^1, \widetilde{\omega}^2, \widetilde{\omega}_1^2$ are expressed by the same equations with the variables x', y', θ' replacing of x, y, θ.

We establish a correspondence between two bivectors by the equation

$$
\widetilde{\omega}^2 = \omega^2.
\tag{6.11}
$$

Since a bivector is defined by three coordinates x, y, θ, and there are only two equations in (6.11), the correspondence is established between families of bivectors.

By setting

$$
\omega^2 = 0, \quad \omega_1^2 = 0,
\tag{6.12}
$$

we see from formulas (6.9) that the vectors $\mathbf{I}_1$ and $\mathbf{I}_2$ are not rotated, and the point M describes the axis x. Thus, equations (6.11) define a family of bivectors with the common axis x.

We see from equations (6.11) that to this family, there are corresponding bivectors with the common axis x' in the second plane. Hence system (6.11) establishes a correspondence of straight lines on two planes.

Substituting expressions (6.10) into equations (6.11), we obtain

$$
\begin{cases}
dx' \sin \theta' + dy' \cos \theta' = -dx \sin \theta + dy \cos \theta, \\
d\theta' = d\theta.
\end{cases}
\tag{6.13}
$$

The integration of the second equation of (6.13) gives

$$\theta' = \theta + c, \quad c = \text{const.},$$

and then the first equation of (6.13) can be written in the form

$$d\left(-x' \sin \theta' + y' \cos \theta'\right) + (x' \cos \theta' + y' \sin \theta')d\theta'$$
$$= d(-x \sin \theta + y \cos \theta) + (-x \cos \theta + y \sin \theta)d\theta.$$

or in the form

$$d\left(-x' \sin \theta' + y' \cos \theta' + x \sin \theta - y \cos \theta\right)$$
$$= (x' \cos \theta' + y' \sin \theta' - x \cos \theta - y \sin \theta)d\theta,$$

i.e, we obtain the equation

$$du = v\,dw.$$

This equation is satisfied by an arbitrary function

$$u = f(w),$$

if one takes

$$\frac{du}{dw} = v, \quad \text{i.e.,} \quad v = f'(w).$$

Thus, the straight line

$$x \cos \theta + y \sin \theta - p = 0 \tag{6.14}$$

is transformed into a new straight line following the law

$$\theta' = \theta + c, \quad p' = p + \dot{f}, \quad \dot{f} = \frac{df}{d\theta},$$

since

$$v = x' \cos \theta' + y' \sin \theta' + p' - (x \cos \theta + y \sin \theta + p)$$
$$= \frac{d}{d\theta}(-x' \sin \theta' + y' \cos \theta' + f + x \sin \theta - y \cos \theta)$$
$$= \frac{du}{d\theta}.$$

The transformation of straight line (6.14) into straight line

$$x \cos (\theta + c) + y \sin (\theta + c) - (p + f) = 0 \tag{6.15}$$

depends on an arbitrary function

$$f = f(\theta)$$

and is not a motion of the plane, although it defines a first-order deformation of the plane considered as a locus of straight lines.

36. Ruled space. A straight line, or it is better to say, a family of trihedrons with a common axis, is defined by equations (5.13):

$$\omega^1 = 0, \quad \omega^2 = 0, \quad \omega_1^3 = 0, \quad \omega_2^3 = 0.$$

Suppose that we have two families (spaces) of trihedrons such that for them

$$\widetilde{\omega}^1 = \omega^1, \quad \widetilde{\omega}^2 = \omega^2, \quad \widetilde{\omega}_1^3 = \omega_1^3, \quad \widetilde{\omega}_2^3 = \omega_2^3. \tag{6.16}$$

Whenever we take a family of trihedrons with a common axis in one of these spaces, the corresponding trihedrons of another space form a family with a common axis too. In other words, formulas (6.16) establish a correspondence between straight lines of these two spaces. Moreover, if we move the second space in such a way that one of its straight lines and one of its trihedrons coincide with the corresponding straight line and trihedron of the first space, then each infinitesimally close straight line of the second space coincides with the corresponding straight line of the first space.

Then the main equations give: for the trihedron T

$$\begin{cases} d\omega^1 = \omega^2 \wedge \omega_2^1 + \omega^3 \wedge \omega_3^1, \\ d\omega^2 = \omega^1 \wedge \omega_1^2 + \omega^3 \wedge \omega_3^2, \\ d\omega_1^3 = \omega_1^2 \wedge \omega_2^3, \\ d\omega_2^3 = \omega_2^1 \wedge \omega_1^3, \end{cases}$$

and after substitution (6.16) for the trihedron T'

$$\begin{cases} d\omega^1 = \omega^2 \wedge \widetilde{\omega}_2^1 + \widetilde{\omega}^3 \wedge \omega_3^1, \\ d\omega^2 = \omega^1 \wedge \widetilde{\omega}_1^2 + \widetilde{\omega}^3 \wedge \omega_3^2, \\ d\omega_1^3 = \widetilde{\omega}_1^2 \wedge \omega_2^3, \\ d\omega_2^3 = \widetilde{\omega}_2^1 \wedge \omega_1^3. \end{cases}$$

From the last two pairs of equations it follows that

$$\widetilde{\omega}_1^2 = \omega_1^2.$$

The first two pairs of equations give

$$(\widetilde{\omega}^3 - \omega^3) \wedge \omega_3^1 = 0, \quad (\widetilde{\omega}^3 - \omega^3) \wedge \omega_3^2 = 0,$$

and this implies that

$$\widetilde{\omega}^3 - \omega^3 = \alpha \omega_1^3, \quad \widetilde{\omega}^3 - \omega^3 = \beta \omega_2^3,$$

where α and β are constants.

If the forms ω_1^3 and ω_2^3 are not linearly dependent, then $\alpha = \beta = 0$, and

$$\widetilde{\omega}^3 = \omega^3,$$

i.e., *the ruled space is rigid.*

Note that if the forms ω_1^3 and ω_2^3 are linearly dependent, then the manifold of straight lines described by the z-axes of trihedrons is not four-dimensional—it is three-dimensional.

Chapter 7

Vector Analysis in an n-Dimensional Euclidean Space

37. Transformation of the space with preservation of a line element.
We return now to the point space. Suppose that it is n-dimensional.

In the Weyl theorem (see No. **32**), for proving space rigidity, we considered separately the components

$$\omega^1, \omega^2, \ldots, \omega^n.$$

We now prove that it is sufficient to consider the sum of squares of these components, i.e., the line element

$$ds^2 = (\omega^1)^2 + (\omega^2)^2 + \ldots + (\omega^n)^2. \tag{7.1}$$

Theorem 1. *Any point correspondence of the space preserving its line element ds^2 reduces to a translation.*

This theorem is stronger than the Weyl theorem. In the latter, we required that when a point M' moves to a point M, every point M_1' that is infinitesimally close to M' moves to the corresponding point M_1 of a neighborhood of M. Now we require only length preservation:

$$MM_1 = M'M_1.$$

To prove this theorem, we first prove the following lemma.

Lemma 2. *The line element ds^2 depends only on the coordinates of a trihedron vertices and their derivatives.*

Proof. Suppose that we have n linearly independent Pfaffian forms

$$\omega^1, \omega^2, \ldots, \omega^n,$$

expressed by means of $r > n$ variables. Suppose that these forms satisfy the conditions

$$d\omega^i = \omega^k \wedge \omega^i_k \qquad (\text{II}_1)$$

and

$$\omega^j_i + \omega^i_j = 0. \qquad (7.2)$$

We prove that the line element

$$ds^2 = (\omega^1)^2 + (\omega^2)^2 + \ldots + (\omega^n)^2$$

can be represented in the form

$$ds^2 = g_{ij} du^i du^j, \quad i, j = 1, \ldots, n.$$

The system of equations

$$\omega^1 = 0, \quad \omega^2 = 0, \quad \ldots, \quad \omega^n = 0 \qquad (7.3)$$

is completely integrable, since according to equations (II_1), the exterior differentials $d\omega^i$, or $\omega^k \wedge \omega^i_k$, vanish by means of the equations of system (7.3).

This system has n first integrals. Denote them by

$$u^1, u^2, \ldots, u^n.$$

They define a position of the vertex M of the trihedron, since for

$$u^1 = \text{const.}, \quad u^2 = \text{const.}, \quad \ldots, \quad u^n = \text{const.},$$

all the forms ω^i vanish, and the point M is fixed. We enlarge these n variables by $r - n$ other variables $u^{n+1}, \ldots, u^r$, and take them as new independent variables

$$u^1, u^2, \ldots, u^n, u^{n+1}, \ldots, u^r.$$

Then

$$ds^2 = g_{ij} du^i du^j, \quad i, j = 1, \ldots, n.$$

We prove that none of the g_{ij} depend on $u^{n+1}, \ldots, u^r$.

Define two symbols of differentiation,

d with respect to changing of the vertex, i.e., with respect to $u^1, \ldots, u^r$; and

δ with respect to changing the variables $u^{n+1}, \ldots, u^r$.

We need to prove that

$$\delta\,(ds^2) = 0.$$

Since $\delta u^i = 0$, we have

$$\delta\,du^i = d\delta^i = 0, \quad i = 1, \ldots, n,$$

and as a result,

$$\omega^i(\delta) = 0. \tag{7.4}$$

Set

$$\omega_i^j(\delta) = e_i^j. \tag{7.5}$$

For the two symbols of differentiation d and δ introduced above, equation (II_1) can be written as a bilinear form

$$\delta\omega^i(d) - d\omega^i(\delta) = \begin{vmatrix} \omega^k(\delta) & \omega_k^i(\delta) \\ \omega^k(d) & \omega_k^i(d). \end{vmatrix}$$

From (7.4) and (7.3), it follows that

$$\delta\omega^i = -e_k^i\omega^k, \tag{7.6}$$

where on the right-hand side we have a regular (not exterior) product.

Differentiating the expression

$$ds^2 = \sum_i (\omega^i)^2$$

by means of the differentiation symbol δ, we find that

$$\delta(ds^2) = \delta\sum_i (\omega^i)^2 = 2\sum_i \omega^i\,\delta\omega^i = -2\sum_{i,k} e_k^i\,\omega^i\omega^k.$$

But from (7.2), the forms e_k^i as well as ω_k^i are skew-symmetric. If we take the last sum with respect to the combinations of i and k, then we obtain

$$\delta(ds^2) = -2\sum_{(i,k)} (e_k^i\omega^i\omega^k + e_i^k\omega^k\omega^i)$$

$$= -2\sum_{(i,k)} (e_k^i + e_i^k)\omega^i\omega^k = 0.$$

Thus,

$$\delta(ds^2) = 0.$$

Geometrically, the formula (7.6),

$$\delta\omega^i = -e_k^i\omega^k,$$

gives the change under a transition from one trihedron to another one with the same vertex; here ω^i are the components of the vector of displacement of the trihedron vertex M, $\delta\omega^i$ is the change of the components ω^i under an infinitesimal transformation of coordinates (a rotation of axes).

Now we prove Theorem 1 (the main theorem of metric geometry).

Proof of Theorem 1. Since ds^2 is a positive definite differential form, it can be always represented as a sum of squares:

$$ds^2 = (\omega^1)^2 + (\omega^2)^2 + \ldots + (\omega^n)^2.$$

Choose an orthogonal frame at the point M in such a way that the quantities $\omega^1, \omega^2, \ldots, \omega^n$ are coordinates of a point $M'(MM' = ds)$ with respect to the chosen frame. We assume at the moment that such a choice is possible,

Choosing a frame at the point $\overline{M}$ of the second space, we obtain that $\overline{\omega}^i = \omega^i$. This proves the theorem.   □

38. Equivalence of reduction of a line element to a sum of squares to the choosing of a frame to be orthogonal. We prove the following lemma which we have used at the end of our proof of Theorem 1.

Lemma 3. *A representation of the quadratic form ds^2 as a sum of squares*

$$ds^2 = (\omega^1)^2 + (\omega^2)^2 + \ldots + (\omega^n)^2.$$

leads to an orthogonal frame.

Proof. The lemma follows from equations (1.10) and (1.12): if $g_{ij} = \delta_{ij}$, then the scalar products of the basis vectors $\mathbf{e}_i$ and $\mathbf{e}_j, j \neq i$, vanish, and the vectors $\mathbf{e}_i$ and $\mathbf{e}_j$ are orthogonal. However, we give a direct proof of this lemma.

Suppose that we have a line element of the form (7.1). We consider the coordinate lines defined by the equations

$$\omega^i = \delta_k^i, \quad \delta_k^i = \begin{cases} 0, & \text{if } i \neq k, \\ 1, & \text{if } i = k, \end{cases} \tag{7.7}$$

where $k = 1, \ldots, n$, and consider the unit vectors of the tangents to these lines. Equation (7.1) gives us the rule of computing of the square of a vector as well as of its norm.

In particular, the vectors $\mathbf{I}_i$ defined by equations (7.7) have the following components:

$$\mathbf{I}_1(1, 0, \ldots, 0), \quad \mathbf{I}_2(0, 1, \ldots, 0), \quad \ldots, \quad \mathbf{I}_n(0, 0, \ldots, 1). \tag{7.8}$$

The scalar product of the vectors $\mathbf{U}$ and $\mathbf{V}$ is defined as the coefficient of 2λ in the decomposition of the linear combination of $\mathbf{U} + \lambda \mathbf{V}$:

$$(\mathbf{U} + \lambda \mathbf{V})^2 = \mathbf{U}^2 + 2\lambda \, \mathbf{U} \cdot \mathbf{V} + \lambda^2 \mathbf{V}^2.$$

Meanwhile, for the vectors (7.8), the linear combination of $\mathbf{I}_1 + \lambda \mathbf{I}_2$ has the coordinates

$$(1, \lambda, \ldots, 0),$$

and from (7.1), its square is

$$(\mathbf{I}_1 + \lambda \mathbf{I}_2)^2 = 1 + \lambda^2.$$

When we find the decomposition of this linear combination directly, we have

$$(\mathbf{I}_1 + \lambda \mathbf{I}_2)^2 = 1 + 2\lambda \, \mathbf{I}_1 \cdot \mathbf{I}_2 + \lambda^2.$$

It follows from the last two equations that $\mathbf{I}_1 \cdot \mathbf{I}_2 = 0$, and the vectors $\mathbf{I}_1$ and $\mathbf{I}_2$ are orthogonal.   □

We can give a *kinematic treatment* of Theorem 1. Define straight lines as lines giving the minimum value for ds^2. Since by the theorem's hypothesis ds^2 is preserved, after transformation these lines are transferred into straight lines. The segments will have their lengths preserved. Suppose that a material system can move, but in such a way that the distances between its points are preserved. Then *the system is not deformable and is transferred as a whole.*

39. Congruence and symmetry. If a line element ds^2 is given, then the space is known. Two spaces with the same ds^2 can be superposed by a simple translation. But there exist two kinds of motions: *motions of the first kind*, or *proper motions*, that after a superposition of figures lead to their *congruence*, and *motions of the second kind*, or *improper motions*, that after a superposition of figures determine their *symmetry*. We need to clarify whether the motions of the first and second kinds are represented by the same analytic expression or, for the left and right trihedrons, whether the components of infinitesimal displacement are different.

Consider a set of points M and a system of right trihedrons with their vertices at M and with components ω^i and ω_i^j given as functions of u^1, u^2, u^3. Consider also a set of points M' and a system of left trihedrons with their vertices at M' and components $\overline{\omega}^i$ and $\overline{\omega}_i^j$ given as functions of v^1, v^2, v^3.

Moving a left trihedron, we can superpose the point M' with the point M and the axes $\mathbf{I}_1, \mathbf{I}_2$ with the axes $\mathbf{I}_1', \mathbf{I}_2'$. However, the axes $\mathbf{I}_3$ and $\mathbf{I}_3'$ will have opposite directions. It follows that

$$\overline{\omega}^1 = \omega^1, \quad \overline{\omega}^2 = \omega^2, \quad \overline{\omega}^3 = -\omega^3.$$

Applying the structure equations

$$d\omega^i = \omega^k \wedge \omega_k^i,$$

we find that

$$\overline{\omega}_1^2 = \omega_1^2, \quad \overline{\omega}_1^3 = -\omega_1^3, \quad \overline{\omega}_2^3 = -\omega_2^3.$$

The system

$$\begin{cases} \overline{\omega}^1(v, dv) = \omega^1(u, du), & \overline{\omega}_1^2(v, dv) = \omega_1^2(u, du), \\ \overline{\omega}^2(v, dv) = \omega^2(u, du), & \overline{\omega}_1^3(v, dv) = -\omega_1^3(u, du), \\ \overline{\omega}^3(v, dv) = -\omega^3(u, du), & \overline{\omega}_2^3(v, dv) = -\omega_2^3(u, du) \end{cases}$$

is completely integrable but it does not produce a motion of the first kind. If for $u^i = u_0^i, v^i = v_0^i$, a point M_0' coincides with a point M_0, then the points

$$M(\omega^1, \omega^2, \omega^3) \text{ and } \overline{M}(\omega^1, \omega^2, -\omega^3)$$

do not coincide—these points are symmetric with respect to the point M_0.

The congruence and the symmetry are determined by different systems of equations.

40. Determination of forms ω_i^j for given forms ω^i. We pose the *problem*: given the forms $\omega^1, \ldots, \omega^n$, find the forms ω_i^j in such a way that they satisfy system (II$_1$) of No. **32** and are skew-symmetric (see (7.2)):

$$d\omega^i = \omega^k \wedge \omega_k^i$$

$$\omega_i^j + \omega_j^i = 0.$$

The exterior differential $d\omega^i$ is a quadratic form with respect to the differentials $du^1, du^2, \ldots, du^n$, and hence with respect to the forms $\omega^1, \omega^2, \ldots, \omega^n$. Thus, this form can be represented as

$$d\omega^i = \sum_{(i,k)} c_{jk}^i \omega^j \wedge \omega^k, \tag{7.9}$$

where the sum is taken over the combinations of indices. In addition, we assume that

$$\omega_i^j = \gamma_i{}^j{}_k \, \omega^k. \tag{7.10}$$

Then equation (7.2) implies that

$$\gamma_{ijk} + \gamma_{jik} = 0, \tag{7.11}$$

and equations (II$_1$) and (7.9) give

$$c_{ijk} = \gamma_{jik} - \gamma_{kij}, \tag{7.12}$$

since after substituting (7.10) into equations (II$_1$) the product $\omega^i \wedge \omega^k$ is obtained from two terms, $\omega^j \wedge \omega_j^i$ and $\omega^k \wedge \omega_k^i$. Next, we apply to equation (7.12) the cyclic permutation of the indices i, j, k twice. This gives

$$\gamma_{jik} - \gamma_{kij} = c_{ijk},$$

$$\gamma_{kji} - \gamma_{ijk} = c_{jki},$$

$$\gamma_{ikj} - \gamma_{jki} = c_{kij}.$$

Subtracting the first two equations from the third one and applying (7.11), we find that

$$\gamma_{ijk} = \frac{c_{kij} - c_{ijk} - c_{jki}}{2}. \tag{7.13}$$

This value of γ_{ijk} satisfies all three equations, and as a result, it satisfies equations (II$_1$).

There is no indication that ds^2 corresponds to Euclidean space.

41. Three-dimensional case. In *the case of three dimensions,* we can proceed differently; in this case, the forms ω_i^j signify infinitesimally small rotations. Set

$$\begin{cases} \omega_2^3 = p_1\omega^1 + p_2\omega^2 + p_3\omega^3, \\ \omega_3^1 = q_1\omega^1 + q_2\omega^2 + q_3\omega^3, \\ \omega_1^2 = r_1\omega^1 + r_2\omega^2 + r_3\omega^3. \end{cases} \qquad (7.14)$$

System (II_1) takes the form

$$\begin{cases} d\omega^1 = \omega^2 \wedge \omega_2^1 + \omega^3 \wedge \omega_3^1 = -(q_2 + r_3)\,\omega^2 \wedge \omega^3 + q_1\,\omega^3 \wedge \omega^1 + r_1\,\omega^1 \wedge \omega^3, \\ d\omega^2 = \omega^1 \wedge \omega_1^2 + \omega^3 \wedge \omega_3^2 = -(p_1 + r_3)\,\omega^3 \wedge \omega^1 + r_2\,\omega^1 \wedge \omega^2 + p_2\,\omega^2 \wedge \omega^3, \\ d\omega^3 = \omega^1 \wedge \omega_1^3 + \omega^2 \wedge \omega_2^3 = -(q_2 + p_1)\,\omega^1 \wedge \omega^2 + p_3\,\omega^2 \wedge \omega^3 + q_3\,\omega^3 \wedge \omega^1. \end{cases}$$
$$(7.15)$$

Apply the above formulas to a line element given in polar coordinates:

$$ds^2 = d\rho^2 + \rho^2 d\theta^2 + \rho^2 \sin^2 \theta d\varphi^2.$$

For the components ω^i, we can take the expressions

$$\omega^1 = d\rho, \quad \omega^2 = \rho d\theta, \quad \omega^3 = \rho \sin \theta d\varphi.$$

Three unit vectors $\mathbf{I}_1, \mathbf{I}_2, \mathbf{I}_3$ correspond to the increments $d\rho, d\theta, d\varphi$. Take the exterior differentials:

$$\begin{cases} d\omega^1 = 0, \\ d\omega^2 = d\rho \wedge d\theta = \dfrac{1}{\rho}\omega^1 \wedge \omega^2, \\ d\omega^3 = \sin \theta\, d\rho \wedge d\varphi + \rho \cos \theta\, d\theta \wedge d\varphi = \dfrac{1}{\rho}\omega^1 \wedge \omega^3 + \dfrac{\cot \theta}{\rho}\omega^2 \wedge \omega^3. \end{cases}$$

Comparing these expressions with (7.15), we find that

$$\begin{cases} q_2 + r_3 = 0, & q_1 = 0, & r_1 = 0, \\ p_1 + r_3 = 0, & r_2 = \dfrac{1}{\rho}, & p_2 = 0, \\ p_1 + q_2 = 0, & p_2 = q_2 = 0, & q_2 = -\dfrac{1}{\rho}, \end{cases}$$

or

$$\begin{cases} p_1 = q_1 = r_1 = 0, & p_2 = q_2 = 0, & r_3 = 0, \\ r_2 = \dfrac{1}{\rho}, & p_3 = \dfrac{\cot \theta}{\rho}, & q_3 = -\dfrac{1}{\rho}. \end{cases}$$

By (7.14), this implies that

$$
\begin{cases}
\omega_2^3 = \dfrac{\cot\theta}{\rho}\omega^3 = \cos\theta\, d\varphi, \\[2mm]
\omega_3^1 = -\dfrac{1}{\rho}\omega^3 = -\sin\theta\, d\varphi, \\[2mm]
\omega_1^2 = \dfrac{1}{\rho}\omega^2 = d\theta.
\end{cases}
$$

Geometrically, it is obvious that the increment $d\varphi$ does not rotate the trihedron, the increment $d\theta$ rotates it about the axis $\mathbf{I}_3$, and the increment $d\varphi$ rotates it about the axis $\mathbf{I}_1\cos\theta - \mathbf{I}_2\sin\theta$.

A line element of a Euclidean space. How we can see for ourselves that ds^2 defines a Euclidean space?

Represent ds^2 as a sum of squares,

$$ds^2 = (\omega^1)^2 + (\omega^2)^2 + \ldots + (\omega^n)^2.$$

As above, we compute the forms ω_i^j and verify whether they satisfy the equations

$$d\omega_i^j = \omega_i^k \wedge \omega_k^j. \tag{7.16}$$

If the answer is positive, then ds^2 belongs to a Euclidean space.

We recommend performing this verification for ds^2 given in polar coordinates.

42. Absolute differentiation. Consider a vector field, i.e., suppose that a vector with components

$$(X^1, X^2, \ldots, X^n)$$

is attached to any point M of the space. When we move from the point M to an infinitesimally close point, the vector undergoes a change (a variation).

We write the vector as

$$\mathbf{X} = X^k \mathbf{I}_k.$$

Its differential is

$$d\mathbf{X} = \mathbf{I}_k dX^k + \omega_k^h X^k \mathbf{I}_h.$$

If we change, in the second sum, the summation indices, k for h and h for k, we obtain

$$d\mathbf{X} = \mathbf{I}_k (dX^k + \omega_h^k X^h).$$

Here dX^k are relative changes of the components of the vector $\mathbf{X}$, and

$$DX^k = dX^k + \omega_h^k X^h. \tag{7.17}$$

are their total changes. This is the *absolute differential* of the components of the vector $\mathbf{X}$. The absolute differential reduces to the true differential if the axes remain parallel.

APPLICATION. *Acceleration of a moving point.* Suppose that

$$v^1, v^2, \ldots, v^n \text{ are components of the velocity,}$$

and
$$\omega^i = v^i dt \text{ are components of a displacement.}$$

The acceleration is the absolute differential of the velocity. The components of the acceleration are defined by the formula

$$g^k = \frac{dv^k}{dt} + \omega^k_h v^h.$$

Suppose that
$$\omega^j_i = \gamma_{ijk}\omega^k.$$

Then
$$g^k = \frac{dv^k}{dt} + v^h \gamma_{hkl} v^l,$$

or after the substitution of indices $\begin{smallmatrix} k\,h\,l \\ l\,k\,h \end{smallmatrix}$,

$$g^i = \frac{dv^i}{dt} + \gamma_{kih}v^k v^h = \frac{dv^i}{dt} - \gamma_{ikh}v^k v^h.$$

When will we have
$$g^i = \frac{dv^i}{dt} \ ?$$

This will be the case if $\gamma_{ikh} + \gamma_{ihk} = 0$, i.e., γ_{ikh} changes its sign if we permute the last two indices. But γ_{ikh} does not change under the cyclic permutation of indices. Thus, all γ_{ikh} vanish, except

$$\gamma_{123} = \gamma_{231} = \gamma_{312} = a.$$

Hence
$$\omega^3_2 = a\omega^1, \quad \omega^1_3 = a\omega^2, \quad \omega^2_1 = a\omega^3.$$

As a result, we have

$$d\omega^3_2 = \omega^2_1 \wedge \omega^1_3 = -a^2 \, \omega^2 \wedge \omega^3,$$
$$d\omega^1 = \omega^2 \wedge \omega^1_2 + \omega^3 \wedge \omega^1_3 = -2a \, \omega^2 \wedge \omega^3,$$

and
$$-a^2 \, \omega^2 \wedge \omega^3 = da \wedge \omega^1 - 2a^2 \, \omega^2 \wedge \omega^3,$$

or
$$da \wedge \omega^1 - 2a^2 \, \omega^2 \wedge \omega^3 = 0.$$

But the product $\omega^2 \wedge \omega^3$ is not divisible by ω^1. Thus, both terms of the last equation vanish separately.

It follows that
$$a = 0,$$

Therefore, all $d\omega^i = 0$, i.e., ω^i are total differentials:

$$\omega^1 = du, \quad \omega^2 = dv, \quad \omega^3 = dw,$$

and

$$ds^2 = du^2 + dv^2 + dw^2.$$

43. Divergence of a vector. Given a closed surface S, suppose that $d\sigma$ is an element of surface of this surface.

Consider the flux of a vector $\mathbf{X}(X^1, X^2, X^3)$ through a surface S. Suppose that d and δ are two symbols of differentiation corresponding to the sides of an infinitesimally small parallelogram for a partition of the surface S. The flux of the vector $\mathbf{X}$ through an element of S is

$$\Omega^1 = \begin{vmatrix} X^1 & X^2 & X^3 \\ \omega^1(d) & \omega^2(d) & \omega^3(d) \\ \omega^1(\delta) & \omega^2(\delta) & \omega^3(\delta) \end{vmatrix} \tag{7.18}$$

$$= X^1 \omega^2 \wedge \omega^3 + X^2 \omega^3 \wedge \omega^1 + X^3 \omega^1 \wedge \omega^2.$$

According to Stokes' formula, we have

$$\iint_S \Omega = \iiint_V d\Omega.$$

The divergence of the vector $\mathbf{X}$, div $\mathbf{X}$, is defined by the formula

$$\iiint_V d\Omega = \iiint_V \operatorname{div} \mathbf{X}\, \omega^1 \wedge \omega^2 \wedge \omega^3. \tag{7.19}$$

i.e., the divergence is the coefficient of the volume element $\omega^1 \wedge \omega^2 \wedge \omega^3$ in the expression for the exterior differential $d\Omega$.

EXAMPLE. *Find the divergence of a vector field in the space referred to the polar coordinates (see No. 41)*

$$\omega^1 = d\rho, \quad \omega^2 = \rho d\theta, \quad \omega^3 = \rho \sin\theta d\varphi.$$

Formula (7.18) has the form

$$\Omega = X^1 \rho^2 \sin\theta\, d\theta \wedge d\varphi + X^2 \rho \sin\theta\, d\varphi \wedge d\rho + X^3 \rho\, d\rho \wedge d\theta,$$

and the exterior differential of the form Ω is

$$d\Omega = \frac{\partial}{\partial\rho}(X^1 \rho^2 \sin\theta)\, d\rho \wedge d\theta \wedge d\varphi + \frac{\partial}{\partial\theta}(X^2 \rho \sin\theta)\, d\rho \wedge d\varphi \wedge d\rho$$

$$+ \frac{\partial}{\partial\varphi}(X^3 \rho)\, d\rho \wedge d\theta \wedge d\varphi = \operatorname{div} \mathbf{X}\, \rho^2 \sin\theta\, d\rho \wedge d\theta \wedge d\varphi.$$

Thus we are able to find div $\mathbf{X}$; *there is no need to know the forms* ω_i^j.

Another method for computing the divergence. Suppose that we have a vector field **X**, and an orthogonal frame is attached to any point of the space. Then from equations (7.17), the absolute differential of **X** is

$$DX^i = dX^i + \omega^i_k X^k.$$

It vanishes when $\omega^1 = \omega^2 = \omega^3 = 0$. It also does not depend on the chosen orientation of a frame. Thus, the absolute differential is a linear form with respect to ω^i:

$$DX^i = X_{i1}\omega^1 + X_{i2}\omega^2 + X_{i3}\omega^3.$$

It is obvious that DX^i does not depend on the choice of a frame.

The quantities X_{ik} define the *derivative tensor* of the vector field $\mathbf{X} = (X^1, X^2, X^3)$ (cf. No. **51**). In order to prove this, it is sufficient to note that the scalar product

$$d\mathbf{X} \cdot \delta\mathbf{M} = DX_i\,\omega^i(\delta) = X_{ij}\,\omega^i(d)\,\omega^j(\delta)$$

is invariant.

Since we now have

$$d(X_1\,\omega^2 \wedge \omega^3 + X_2\,\omega^3 \wedge \omega^1 + X_3\,\omega^1 \wedge \omega^2)$$
$$= (X_{11} + X_{22} + X_{33})\,\omega^1 \wedge \omega^2 \wedge \omega^3,$$

it follows that

$$\operatorname{div}\mathbf{X} = X_{11} + X_{22} + X_{33}, \tag{7.20}$$

i.e., the divergence is the contracted tensor of the derivative tensor of a vector field.

44. Differential parameters. Let V be a scalar function of class C^2 and

$$dV = V_1\omega^1 + \ldots + V_n\omega^n.$$

The quantities V_i define a vector called the *gradient* of the function V. We have

$$dV = \operatorname{grad}\mathbf{V} \cdot d\mathbf{M} = V_i\,\omega^i.$$

In orthogonal Cartesian coordinates, the quantities V_i are expressed as follows:

$$V_1 = \frac{\partial V}{\partial u_1}, \quad V_2 = \frac{\partial V}{\partial u_2}, \quad \ldots, \quad V_n = \frac{\partial V}{\partial u_n}.$$

The *first differential parameter* is a square of the norm of the gradient:

$$\Delta_1 V = (V_1)^2 + \ldots + (V_n)^2. \tag{7.21}$$

The *second differential parameter* is the divergence of the gradient:

$$\Delta_2 V = V_{11} + V_{22} + \ldots + V_{nn}. \tag{7.22}$$

In orthogonal Cartesian coordinates, the second differential parameter is

$$\Delta_2 V = \frac{\partial^2 V}{(\partial x^1)^2} + \frac{\partial^2 V}{(\partial x^2)^2} + \ldots + \frac{\partial^2 V}{(\partial x^n)^2}. \qquad (7.23)$$

In arbitrary coordinates, this parameter is defined by the equation

$$\iint_S (V_1 \, \omega^2 \wedge \omega^3 + V_2 \, \omega^3 \wedge \omega^1 + V_3 \, \omega^1 \wedge \omega^2)$$

$$= \iiint_V \Delta_2 V \, \omega^1 \wedge \omega^2 \wedge \omega^3. \qquad (7.24)$$

EXAMPLE. Compute $\Delta_2 V$ in polar coordinates. We have (see No. **41**)

$$\omega^1 = d\rho, \quad \omega^2 = \rho \, d\theta, \quad \omega^3 = \rho \sin\theta \, d\varphi,$$

$$dV = \frac{\partial V}{\partial \rho} + \frac{1}{\rho} \frac{\partial V}{\partial \theta} \rho \, d\theta + \frac{1}{\rho \sin\theta} \frac{\partial V}{\partial \rho} \rho \sin\theta \, d\varphi.$$

Thus,

$$V_1 = \frac{\partial V}{\partial \rho}, \quad V_2 = \frac{1}{\rho} \frac{\partial V}{\partial \theta}, \quad V_3 = \frac{1}{\rho \sin\theta} \frac{\partial V}{\partial \rho}$$

and the flux of a vector field is

$$\frac{\partial V}{\partial \rho} \rho^2 \sin\theta \, d\theta \wedge d\varphi + \frac{\partial V}{\partial \theta} \sin\theta \, d\varphi \wedge d\rho + \frac{\partial V}{\partial \rho} \frac{1}{\sin\theta} d\varphi \wedge d\theta.$$

Taking the exterior differential and applying formula (7.24), we find that

$$\Delta_2 V \rho^2 \sin\theta \, d\rho \wedge d\theta \wedge d\varphi = \sin\theta \frac{\partial}{\partial \rho} \left(\frac{\partial V}{\partial \rho} \rho^2 \right) d\rho \wedge d\theta \wedge d\varphi$$

$$+ \frac{\partial}{\partial \theta} \left(\sin\theta \frac{\partial}{\partial \theta} \right) d\rho \wedge d\theta \wedge d\varphi + \frac{1}{\sin\theta} \frac{\partial^2 V}{\partial \varphi^2} d\rho \wedge d\theta \wedge d\varphi$$

and

$$\Delta_2 V = \frac{1}{\rho^2} \frac{\partial}{\partial \rho} \left(\frac{\partial V}{\partial \rho} \rho^2 \right) + \frac{1}{\rho^2 \sin^2\theta} \frac{\partial}{\partial \theta} \left(\sin\theta \frac{\partial}{\partial \theta} \right) + \frac{1}{\rho^2 \sin^2\theta} \frac{\partial^2 V}{\partial \varphi^2}.$$

If V depends only on ρ, then

$$\Delta_2 V = \frac{1}{\rho^2} \frac{\partial}{\partial \rho} \left(\rho^2 \frac{\partial V}{\partial \rho} \right).$$

Chapter 8

The Fundamental Principles of Tensor Algebra

45. Notion of a tensor. A *tensor* is a set of numbers attached to a point of the space and defining a certain (geometric or physical) object at this point. Thus, the tensor itself depends on a point, but not on a chosen coordinate system (a frame) although tensor components are changed under the change of a coordinate system (choosing a frame). However, the notion of a tensor is significantly narrower than this definition. As we will see below, the tensor components are changed according to a certain law of transformation.

An example of a tensor is a vector (components of a vector). We saw (see formulas (1.15), (1.16),(1.17),(1.18), and (1.19) in No. **6**) that with respect to a nonorthogonal frame, a vector $\mathbf{X}$ has two series of components, contravariant X^k and covariant X_k; the scalar square of a vector $\mathbf{X}$ is obtained by the contraction of covariant components with contravariant ones:

$$X_i X^i = X_1 X^1 + X_2 X^2 + \ldots + X_n X^n = \mathbf{X}^2,$$

and the scalar product of vectors $\mathbf{X}$ and $\mathbf{Y}$ is obtained by the contraction of contravariant components of one factor with covariant components of another:

$$\mathbf{X} \cdot \mathbf{Y} = X^i Y_i.$$

All these are extended to tensors, and the last formula after such an extension becomes the definition of a tensor. The simplest way to define a tensor rigorously is to request that after its contraction with a sufficient number of covariant or contravariant components of (arbitrary) vectors we obtain a scalar which is invariant with respect to the choice of frame. For example, components a_{ijk} define a triply-covariant tensor if, for contravariant components of three arbitrary vectors $\mathbf{X}, \mathbf{Y}$, and $\mathbf{Z}$, the contracted product

$$a_{ijk} X^i Y^j Z^k \tag{8.1}$$

is an invariant (i.e., it does not depend on the choice of a frame).

Now we can derive the law of infinitesimal transformations of the tensor under a frame variation—the law that can be taken as a tensor definition.

From (7.17), if the initial point of a vector is fixed, then the absolute differential of components of a contravariant vector vanishes,

$$DX^k = dX^k + \omega_h^k X^h = 0. \tag{8.2}$$

Denote a frame variation by the symbol δ and the form $\omega_h^k(\delta)$ by e_h^k. Then it follows from equations (8.2) that[1]

$$\delta X^k = -e_h^{.k} X^h = e_{.h}^k X^h. \tag{8.3}$$

This defines the change of components of a vector under a frame variation. For a triply-covariant tensor a_{ijk}, we have found invariant (8.1). Thus, we have the equation

$$\delta(a_{ijk} X^i Y^j Z^k) = 0,$$

or

$$\delta a_{ijk} X^i Y^j Z^k - a_{ijk} Y^j Z^k e_h^{.i} X^h - a_{ijk} X^i Z^k e_h^{.j} Y^h - a_{ijk} X^i Y^j e_h^{.k} Z^h = 0.$$

If in the last equation we make the substitutions $\binom{h}{i}$, $\binom{h}{j}$, and $\binom{h}{k}$ of indices in the second, third, and fourth sums, respectively, then we have

$$(\delta a_{ijk} - a_{hjk} e_i^{.h} - a_{ihk} e_i^{.h} - a_{ijh} e_k^{.h}) X^i Y^j Z^k = 0.$$

Since the vectors X^i, Y^j, and Z^k are arbitrary, the expression in parentheses vanishes,

$$\delta a_{ijk} = a_{hjk} e_i^{.h} + a_{ihk} e_i^{.h} + a_{ijh} e_k^{.h}. \tag{8.4}$$

This is a necessary and sufficient condition for the quantities a_{ijk} to define a triply-covariant tensor.

The formulas for variations of contravariant tensors have a similar form. Differentiating equations (1.16) of No. **6**,

$$X_i = g_{ij} X^j,$$

by means of the symbol δ (a frame variation) and then applying (8.3) and (5.4), we find that

$$\delta X_i = X^j \delta g_{ij} + g_{ij} \delta X^j = X^j (g_{kj} e_i^{.k} + g_{ik} e_j^{.k}) + g_{ij} e_{.h}^i X^h.$$

Changing the indices $\binom{i\,k}{h\,j}$ in the last sum and applying equations (1.16), we obtain

$$\delta X_i = e_i^{.k} X_k. \tag{8.5}$$

---

[1]In the formula below and in what follows, occasionally Cartan uses the dots or leaves a space in order to emphasize the order of indices. (*Translator's note.*)

Now for a second-order tensor $a_i^{\cdot j}$, one time covariant and one time contravariant, we compose an invariant product $a_i^{\cdot j} X^i Y_j$. This product satisfies the equation

$$\delta(a_i^{\cdot j} X^i Y_j) = 0.$$

Using (8.3) and (8.5), we find from this equation that

$$\delta a_i^{\cdot j} X^i Y_j + a_i^{\cdot j}(e_{\cdot h}^i X^h Y_j + e_j^{\cdot h} X^i Y_k) = 0$$

and

$$\delta a_i^{\cdot j} + a_h^{\cdot j} e_{\cdot i}^h + a_i^{\cdot j} e_h^{\cdot j} = 0,$$

or

$$\delta a_i^{\cdot j} = a_h^{\cdot j} e_i^{\cdot h} + a_i^{\cdot h} e_{\cdot h}^j.$$

In an orthogonal frame, contravariant and covariant components coincide, and the lower or upper location of indices is only to indicate summation.

46. Tensor algebra. 1°. SATURATION OF INDICES. From a tensor a_{ij} with two indices, we can get a *contracted* tensor (a scalar invariant)

$$a_{11} + a_{22} + \ldots + a_{nn}.$$

We shall say that this contracted tensor is obtained from a_{ij} by *saturation* of the indices i and j.

This contracted tensor does not depend on the choice of axes. In fact, changing the indices $\binom{i\,k}{k\,i}$ in the second sum, we obtain

$$\delta \sum_i a_{ii} = -\sum_{(i,k)}(e_i^{\cdot k} a_{ki} + e_i^{\cdot k} a_{ik}) = -\sum_{(i,k)}(e_i^{\cdot k}(a_{ii} + a_{ki})$$

$$= -\sum_{(i,k)} a_{ki}(e_i^{\cdot k} + e_k^{\cdot i}) = 0,$$

since

$$e_i^{\cdot k} + e_k^{\cdot i} = 0.$$

Another proof. The expressions

$$a_{ij} X^i Y^j \quad \text{and} \quad a_{ij} X^j Y^i$$

do not depend on the choice of axes. Thus, their sum

$$a_{ij}(X^i Y^j + X^j Y^i),$$

also does not depend on the choice of axes. If, in the second sum, we change the indices $\binom{ij}{ji}$ and set $Y^i = X^i$, we find that

$$(a_{ij} + a_{ji}) X^i Y^j = b_{ij} X^i Y^j.$$

This is a quadratic symmetric form, since $b_{ij} = a_{ij} + a_{ji}$. It is not difficult to see that $\sum_i b_{ii}$ is an invariant. In fact, consider the difference

$$\sum_{i,j} b_{ij} X^i X^j - \lambda \sum_i (X_i)^2.$$

The vanishing of the discriminant (the characteristic equation)

$$\begin{vmatrix} b_{11} - \lambda & b_{12} & \cdots & b_{1n} \\ b_{21} & b_{22} - \lambda & \cdots & b_{2n} \\ \cdots\cdots\cdots\cdots\cdots\cdots\cdots\cdots \\ b_{n1} & b_{n2} & \cdots & b_{nn} - \lambda \end{vmatrix} = 0$$

does not depend on the choice of axes.

The roots λ of the characteristic equation are scalar tensors, and symmetric functions of roots are invariants. In particular, the sum of roots

$$b_{11} + b_{22} + \ldots + b_{nn} = 2 \sum_i a_{ii}$$

is an invariant.   □

2°. Addition of tensors. Two tensors with the same number of indices produce a tensor if we add their corresponding components:

$$a_{ij} + b_{ij} = c_{ij}.$$

In fact, if a_{ij} and b_{ij} are tensors, then contracting each of them with arbitrary vectors X^i and Y^j, we obtain an invariant. As a result, their sum

$$c_{ij} X^i Y^j = a_{ij} X^i Y^j + b_{ij} X^i Y^j.$$

is also an invariant.

Application. *Any second-order tensor a_{ij} can be represented as a sum of a symmetric and a skew-symmetric tensor.* In fact,

$$b_{ij} = a_{ji}$$

is a tensor, since $b_{ij} X^i Y^j = a_{ji} X^i Y^j$ is a number that does not depend on the choice of axes. It follows that

$$\alpha_{ij} = \frac{1}{2}(a_{ij} + a_{ji}) \text{ is a symmetric tensor,}$$

$$\beta_{ij} = \frac{1}{2}(a_{ij} - a_{ji}) \text{ is a skew-symmetric tensor,}$$

and

$$a_{ij} = \alpha_{ij} + \beta_{ij}.$$

3°. MULTIPLICATION OF TENSORS. Consider first a *general (outer) multiplication*. If a_{ij} and b_{hkl} are tensors, then

$$c_{ijkhl} = a_{ij}b_{khl}$$

is a tensor, since multiplying the last equation by arbitrary vectors X^i, Y^j, Z^k, U^h, and V^l, we obtain the product of invariants which is itself an invariant:

$$c_{ijkhl}X^iY^jZ^kU^hV^l = (a_{ij}X^iY^j)(b_{khl}Z^kU^hV^l).$$

Next we consider a *contracted (inner) multiplication* (or multiplication with contraction). The sum of the products

$$b_{ij} = a_{ijk}X^k$$

is a second-order tensor, since

$$b_{ij}Y^iZ^j = a_{ijk}X^kY^iZ^j$$

is an invariant. This implies that

$$c_k = a_{i.k}^{i}$$

is a first-order tensor. Each contraction reduces the order (the number of indices) by two units.

Next, we discuss a *skew-symmetric tensor*. Consider two vectors X^i and Y^j and their general product

$$a^{ij} = X^iY^j.$$

Compose the difference of these two tensors:

$$c^{ij} = a^{ij} - a^{ji} = X^iY^j - X^jY^i.$$

This difference is a skew-symmetric tensor with $\frac{1}{2}n(n-1)$ components. It is obvious that $c^{ji} = -c^{ij}$.

47. Geometric meaning of a skew-symmetric tensor. We saw that a skew-symmetric tensor is defined by two arbitrary vectors taken in a certain order (a bivector). In order to find its geometric meaning, we clarify when two such tensors are equal.

1°. When does a vector **Z** lie in the plane **(X, Y)** (or a parallel plane) defined by vectors **X** and **Y** ? Consider an equation expressing this condition.

If the matrix

$$\begin{Vmatrix} X^1 & X^2 & \cdots & X^n \\ Y^1 & Y^2 & \cdots & Y^n \\ Z^1 & Z^2 & \cdots & Z^n \end{Vmatrix}$$

has all third-order minors vanishing, then the components Z^i satisfy the system of equations

$$Z^1 c^{23} + Z^2 c^{31} + Z^3 c^{12} = 0,$$
$$\qquad c^{ij} = \begin{vmatrix} X^i & X^j \\ Y^i & Y^j \end{vmatrix}. \qquad (8.6)$$
$$Z^1 c^{24} + Z^2 c^{41} + Z^4 c^{12} = 0,$$

It is obvious that a location of the plane **XY** depends on the bivector components c^{ij} only.

The plane of vectors **X** and **Y** is a significant element of the notion of the tensor c^{ij}. For two skew-symmetric tensors c^{ij} and c_*^{ij} to be equal, it is first necessary that the planes of vectors $(\mathbf{X}, \mathbf{Y})$ and $(\mathbf{X}_*, \mathbf{Y}_*)$ coincide.

2°. Take the plane of the bivector c^{ij} as the plane $(1, 2)$. Then our vectors **X** and **Y** have only two nonvanishing coordinates:

$$\mathbf{X} = X^1 \mathbf{I}_1 + X^2 \mathbf{I}_2,$$
$$\mathbf{Y} = Y^1 \mathbf{I}_1 + Y^2 \mathbf{I}_2,$$

there is only one nonvanishing component of the tensor c^{ij}:

$$c^{12} = X^1 Y^2 - X^2 Y^1,$$

and its remaining components are equal to 0.

It is not difficult to note that this nonvanishing component is equal to the norm of the vector product of the vectors **X** and **Y**:

$$\mathbf{X} \times \mathbf{Y} = (X^1 \mathbf{I}_1 + X^2 \mathbf{I}_2) \times (Y^1 \mathbf{I}_1 + Y^2 \mathbf{I}_2) = (X^1 Y^2 - X^2 Y^1) \mathbf{I}_3.$$

Thus, this is the area of the parallelogram built on the vectors **X** and **Y**. Moreover, the component c^{12} is positive if the rotation from the vector **X** to the vector **Y** is positive (with respect to the chosen frame $\mathbf{I}_1 \mathbf{I}_2 \mathbf{I}_3$). Thus, the bivector c^{ij} is an oriented parallelogram.

Two bivectors are equal if they define the same (or parallel) plane, the same area, and the same location of vectors (the same orientation).

Note that the area of the parallelogram can be replaced by any other oriented area.

EXAMPLE OF A BIVECTOR: an exterior product $\omega^i \wedge \omega^j$ is a bivector representing an element of area, $\omega^i(d)\omega^j(\delta) - \omega^i(\delta)\omega^j(d)$. The sum of two bivectors is not always a bivector. Any skew-symmetric tensor can be considered as a sum (a system) of bivectors.

In a three-dimensional space a skew-symmetric tensor is always a bivector.

Consider a four-dimensional space. If we exclude Z^1 from the first two equations of (8.6), then we obtain

$$Z^2(c^{24} c^{31} - c^{23} c^{41}) + Z^3 c^{24} c^{12} - Z^4 c^{23} c^{12} = 0.$$

But in system (8.6), there is an equation for Z^2, Z^3, Z^4 of the form

$$Z^2 c^{34} + Z^3 c^{12} + Z^4 c^{23} = 0.$$

Multiplying this equation by c^{12} and adding it to the previous equation, we note that not only do the terms with Z^4 cancel out, but so too do the terms with Z^3, since $c^{24} = -c^{42}$. Because now we can assume that $Z^2 \neq 0$, we find that

$$c^{24} c^{31} - c^{23} c^{41} + c^{34} c^{12} = 0. \tag{8.7}$$

It is easy to check that under condition (8.7), system (8.6) is consistent and defines the vectors $\mathbf{Z}$ belonging to the plane $(\mathbf{X}, \mathbf{Y})$. This proves the existence of a plane defined by the tensor c^{ij}, and hence of a pair of vectors $\mathbf{X}, \mathbf{Y}$ defining the bivector c^{ij}.

Therefore, equation (8.7) is a condition for c^{ij} to be a bivector in a four-dimensional space.

48. Scalar product of a bivector and a vector and of two bivectors. PRODUCT OF A BIVECTOR AND A VECTOR. If (a_{ij}) is a bivector, then the contracted product

$$b_i = a_{ji} X^j$$

is a vector.

In order to interpret the vector b_i geometrically, we choose the plane of the bivector (a_{ij}) as the plane (1,2). Then $a_{12} \neq 0$, the remaining $a_{ij} = 0$, and

$$b_1 = -a_{12} X^2, \quad b_2 = a_{12} X^1, \quad b_3 = \ldots = b_n = 0.$$

It is obvious that the vector (b_i) lies in the plane of the bivector (a_{ij}). It depends on the projection of the vector $\mathbf{X}$ on the plane of the bivector: this projection is multiplied by the area of the bivector and is rotated through a right angle.

Thus, *the product (b_i) is the projection of the vector $\mathbf{X}$ on the plane of the bivector (a_{ij}) which is rotated through a right angle in the positive direction of the bivector (a_{ij}) and is multiplied by the area of the bivector.*

TWICE CONTRACTED PRODUCT OF TWO BIVECTORS:

$$a_{ij} b^{ij} = a_{11} b^{11} + 2a^{12} b^{12} + \ldots . \tag{8.8}$$

This is a scalar product. In order to interpret this geometrically, we take the plane of the first bivector (a_{ij}) as the plane (1,2). Then for the first bivector, we have $a_{12} \neq 0$, and the remaining $a_{ij} = 0$. Product (8.8) has the only one term

$$\frac{1}{2} \sum_{(i,j)} a_{ij} b^{ij} = a_{12} b^{12}.$$

This is the product of the area of the first bivector and the area of the orthogonal projection of the second one onto the plane of the first.

Thus, the scalar square of the bivector is

$$(a_{12})^2 + (a_{13})^2 + \ldots , \tag{8.9}$$

and the cosine of the angle between the planes of two bivectors is

$$\cos\varphi = \frac{a_{ij}b^{ij}}{\sqrt{\sum\limits_{i,j}(a_{ij})^2 \sum\limits_{i,j}(b_{ij})^2}}. \tag{8.10}$$

This can be extended to a trivector, etc.

49. Simple rotation of a rigid body around a point. Consider a rotation (ω_i^j) of a frame about the point O. The motion of the translation of a point $P(X^i)$ is

$$X^k\omega_k^i.$$

If we set

$$a_{ki} = \frac{\omega_k^i}{dt},$$

then the velocity v_i of the translation of the point P is

$$v_i = X^k a_{ki}.$$

This introduces a skew-symmetric system of numbers (a_{ki}), since $a_{ij} = -a_{ji}$. This system is a tensor. In fact,

$$a_{ki}X^kY^i = v_iY^i = \mathbf{V}\cdot\mathbf{Y}$$

is a scalar product of two vectors, i.e., it is an invariant. Thus, in a three-dimensional space, a rotation can be represented by a bivector.

Set

$$v_1 = a_{21}X^2 + a_{31}X^3,$$
$$v_2 = a_{12}X^1 + a_{32}X^3,$$
$$v_3 = a_{13}X^1 + a_{23}X^2$$

and introduce the notations

$$a_{23} = p, \quad a_{31} = q, \quad a_{12} = r$$

and

$$X^1 = x, \quad X^2 = y, \quad X^3 = z.$$

Then we obtain the well-known formulas for components of the velocity $\mathbf{V}$ with respect to the coordinate axes:

$$V_x = qz - ry, \quad V_y = rx - pz, \quad V_z = py - qz.$$

For $n > 3$, a *simple rotation* is a rotation which can be analytically represented by a bivector. A general rotation can be represented by a system of bivectors. An arbitrary rotation in the space is a sum of $\frac{1}{2}n(n-1)$ simple rotations having only one component a_{ki}. So,

$$a_{ki}X^k$$

is the contracted product of the bivector (a_{ki}) and the vector (X^k).

Chapter 9

Tensor Analysis

50. Absolute differentiation. We consider a tensor field and find its absolute differential. Let us start with vectors.

1°. VECTORS. The *absolute differential* of a vector field $\mathbf{X}$ is the principal linear part of the difference of two infinitesimally close vectors $\mathbf{X}'$ and $\mathbf{X}$:

$$D\mathbf{X} = \mathbf{X}' - \mathbf{X}.$$

If the axes of frames are fixed (parallel), then the coordinates of the above difference are the differentials

$$dX^i.$$

If the axes move, then we proved in No. **42** that the absolute differential has form (7.17),

$$DX^i = dX^i + \omega_k^i X^k.$$

2°. TENSORS. A similar construction can be done for tensors. Move from a point M to a point M'. If the axes of frames are parallel, then the change of a tensor a_{ij} is determined by the differentials

$$(da_{ij}).$$

If the axes are not parallel, then the change of a tensor a_{ij} is determined by the absolute differential

$$(Da_{ij}).$$

Construct two uniform vector fields X^i and Y^i, i.e., vector fields satisfying the equations

$$\begin{cases} dX^i + \omega_k^i X^k = 0, \\ dY^i + \omega_k^i Y^k = 0. \end{cases} \tag{9.1}$$

By the tensor definition, the contracted inner product

$$c = a_{ij} X^i Y^j$$

75

is an invariant and does not depend on the choice of axes.

We compute the differential dc by two methods: first, when the axes are parallel, and second, when they are not parallel.

In the first case, we have

$$dc = Da_{ij}X^iY^j, \tag{9.2}$$

since the vectors X^i and Y^i are unchanged.

In the second case, we have

$$dc = da_{ij}X^iY^j + a_{ij}dX^iY^j + a_{ij}X^idY^j,$$

or, replacing dX^i and dY^j by their values from formulas (9.1), we find that

$$dc = da_{ij}X^iY^j - a_{ij}\omega_k^iX^kY^j - a_{ij}X^i\omega_k^jY^k.$$

Changing the indices $\binom{i\,k}{k\,i}$ in the second sum and $\binom{j\,k}{k\,j}$ in the third sum, we obtain

$$dc = (da_{ij} - a_{kj}\omega_k^i - a_{ik}\omega_j^k)X^iY^j. \tag{9.3}$$

But c does not depend on the choice of axes. So, dc does not depend on the choice of axes either. Comparing expression (9.2) with (9.3) that we have obtained for dc, we get

$$Da_{ij} = da_{ij} - a_{kj}\omega_k^i - a_{ik}\omega_j^k. \tag{9.4}$$

This is the *absolute differential of a tensor*.

51. Rules of absolute differentiation. 1°. Differential of a general product.

Theorem 1. *If*

$$c_{ijk} = a_ib_{jk},$$

then

$$Dc_{ijk} = (Da_i)b_{jk} + a_i(Db_{jk}). \tag{9.5}$$

Proof. The theorem is obvious if the axes of frames at points M and M' are parallel. Since absolute differentiation does not depend on the choice of frames, formula (9.5) is always valid. □

2°. Differential of a contracted product.

Theorem 2. *If*

$$c_i = a_kb_i^{\cdot k},$$

then

$$Dc_k = (Da_k)b_i^{\cdot k} + a_k(Db_i^{\cdot k}). \tag{9.6}$$

Proof. The proof is the same as for Theorem 1.   □

3°. DERIVATIVE TENSOR OF A GIVEN TENSOR. Absolute differentiation allows us to associate with a given tensor field a new tensor field, whose order is one unit greater.

The absolute differential vanishes if we remain at the same point M, i.e., if $\omega^k = 0$. Since Da_{ij} vanishes if $\omega^k = 0$, the absolute differential Da_{ij} is a linear function of ω^k:

$$Da_{ij} = a_{ijk}\omega^k. \tag{9.7}$$

We prove that the quantities a_{ijk} define a new third-order tensor.

Denote

$$b_{ij} = a_{ijk}X^k,$$

where X^k is a vector, for example, the velocity vector of a moving point passing through the point M, whose components are ω^k. If we denote the time by t, then the components of the velocity $\mathbf{X}$ are

$$X^k = \frac{\omega^k}{dt}.$$

If we divide both parts of this equation by dt, then we obtain

$$\frac{Da_{ij}}{dt} = a_{ijk}\frac{\omega^k}{dt} = a_{ijk}X^k = b_{ij}.$$

The left-hand side of the last equation is expressed in terms of the absolute differential and defines the rate of change of the component a_{ij}, when we move along a curve for which the vector $\mathbf{X}$ (the velocity) is a tangent. Thus, b_{ij} is a second-order tensor, since the derivative of the tensor a_{ij} in a given direction is an absolute derivative (i.e., a derivative not depending on the choice of axes).

Thus, for any vectors $\mathbf{Y}$ and $\mathbf{Z}$, the contracted product

$$b_{ij}Y^iZ^j = a_{ijk}X^kY^iZ^j$$

is an invariant, and as a result, a_{ijk} is a tensor. This tensor is called the *derivative tensor* of the tensor a_{ij}.

EXAMPLE. Suppose that V is any C^2-function of a point. Then V^i is the gradient (i.e., a vector), V^{ij} is a second-order tensor, etc.

52. Exterior differential tensor-valued form. We begin with a scalar form. If a_i is a vector, then

$$\mathbf{a} \cdot d\mathbf{M} = a_i\omega^i$$

is a form which does not depend on the choice of axes, since it is the scalar product of the vector a_i and the vector ω^i.

For a bivector a_{ij}, the product $a_{ij}\,\omega^i \wedge \omega^j$ is an invariant exterior differential form provided that $a_{ij} = -a_{ji}$. In fact, $\omega^i \wedge \omega^j$ is a surface element, i.e, it is a bivector, and $a_{ij}\,\omega^i \wedge \omega^j$ is the scalar product of two bivectors.

This can be generalized. If a tensor a_{ijk} satisfies the conditions

$$a_{ijk} = a_{jki} = a_{kij} = -a_{ikj}, \text{ etc.}$$

i.e., it defines a trivector, then the form

$$a_{ijk}\omega^i \wedge \omega^j \wedge \omega^k$$

does not depend on the choice of axes, since it is the scalar product of two trivectors.

The *vector-valued forms* can be constructed in a similar way. Let a_{ij} be a tensor. Then

$$\widetilde{\omega}_i = a_{ij}\omega^j$$

is a vector attached to a pair of infinitesimally close points M and $M+d M$. This vector is the contracted product of the tensor a_{ij} and an infinitesimal vector of the given direction ω^k.

Similarly, if a_{ijk} is a tensor, which is skew-symmetric with respect to the indices j and k, i.e., if

$$a_{ijk} = -a_{ikj},$$

then

$$\omega_i = a_{ijk}\omega^j \wedge \omega^k$$

is a vector referred to an area element, or a *vector-valued exterior quadratic form*.

Further, we construct a tensor-valued form. If a_{ijk} is a tensor which is skew-symmetric with respect to the first two indices i and j,

$$a_{ijk} = -a_{jik},$$

then the form

$$\widetilde{\omega}_{ij} = -a_{ijk}\omega^k$$

is a skew-symmetric second-order tensor referred to a line element defined by the components ω^k, or a bivector-valued Pfaffian form.

78. A problem of absolute exterior differentiation. Consider a linear differential form $\widetilde{\omega}^i$ and a volume V bounded by a surface S. Suppose that a vector $\widetilde{\omega}^i$ is attached to any point of S. The geometric sum of vectors (vector-valued forms) on the surface S is equal to the geometric sum of the exterior differentials of $\widetilde{\omega}^i$ applied to all points of the volume V. If the axes are parallel at all points of the region, then by Stokes' formula (4.8), we have

$$\iint_S \widetilde{\omega}^i = \iiint_V d\widetilde{\omega}^i.$$

If the axes are not parallel, then we introduce a uniform vector field X_i and consider the sum of the products

$$X_i\widetilde{\omega}^i.$$

The exterior differential does not depend on the choice of axes. If the axes are parallel, then the exterior differential is

$$X_i \, d\widetilde{\omega}^i,$$

since the field is uniform, the field X_i is not differentiated. If the axes are not parallel, then the exterior differential has the form

$$X_i \, d\widetilde{\omega}_i + dX_i \wedge \widetilde{\omega}^i.$$

Applying formulas (8.2),

$$dX_i = \omega_i^k X_k,$$

we can write it as

$$X_i \, d\widetilde{\omega}^i + X_k \omega_i^k \wedge \widetilde{\omega}^i = X_i \, d\widetilde{\omega}^i + X_i \omega_k^i \wedge \widetilde{\omega}^k.$$

Denote the absolute differential of the form $\widetilde{\omega}^i$ by $D\widetilde{\omega}^i$. Then we have

$$X_i \, D\widetilde{\omega}^i = X_i(d\widetilde{\omega}^i + \omega_k^i \wedge \widetilde{\omega}^k),$$

and this implies that

$$D\widetilde{\omega}^i = d\widetilde{\omega}^i + \omega_k^i \wedge \widetilde{\omega}^k. \tag{9.8}$$

We have found the same rule as before.

EXAMPLE. Obviously, $\widetilde{\omega}^i = \omega^i$ is a tensor.

Compute the absolute exterior differential $D\omega^i$:

$$D\omega^i = d\omega^i + \omega_k^i \wedge \omega^k = d\omega^i - \omega^k \wedge \omega_k^i = 0.$$

It is equal to 0 by means of the main equations. It is easy to understand: if the axes are orthogonal, then $\omega^i = dx^i$, i.e., ω^i are exact differentials.

The tensor algebra is extended to differential forms. If the forms ω^i and θ_i are given, then the exterior product

$$\widetilde{\omega}^i \wedge \theta_j$$

is also a tensor. For example, $\omega^i \wedge \omega^k$ is a surface element.

Theorem 3. *For any given two tensors $\widetilde{\omega}_{ij}$ and θ_k, the exterior differential of their exterior product $\widetilde{\omega}_{ij} \wedge \theta_k$ is*

$$D\widetilde{\omega}_{ij} \wedge \theta_k + (-1)^p \widetilde{\omega}_{ij} \wedge D\theta_k, \tag{9.9}$$

where $D\widetilde{\omega}_{ij}$ is the absolute exterior differential of the form $\widetilde{\omega}_{ij}$, $D\theta_k$ is the absolute exterior differential of the form θ_k, and p is the order of the form $\widetilde{\omega}_{ij}$ (i.e., the number of differentials that are multiplied in the expression of $\widetilde{\omega}_{ij}$).

For the proof, it is sufficient to take parallel axes.

EXAMPLE:

$$\omega^i \wedge \omega^j.$$

The exterior differential of the product $\omega^i \wedge \omega^j$ vanishes,

$$D\omega^i \wedge \omega^j - \omega^i \wedge D\omega^j = 0,$$

since

$$D\omega^i = D\omega^j = 0.$$

B. THE THEORY OF RIEMANNIAN MANIFOLDS

Chapter 10

The Notion of a Manifold

54. The general notion of a manifold. It is rather difficult to define precisely the general notion of a manifold. A surface gives some idea of a two-dimensional manifold. If we take a sphere or a torus, we can divide each of these surfaces into a finite number of regions such that there exists a one-to-one continuous representation of each of these regions on a simply connected region of the Euclidean plane.

More precisely, given an arbitrary point P_0 on the manifold, it is possible to find in a neighborhood of the point P_0, a coordinate system u, v such that if u_0, v_0 are the coordinates of the point P_0, then there exists a positive number r having the following property. Every system of numbers u, v satisfying the inequality

$$(u - u_0)^2 + (v - v_0)^2 < r^2 \tag{10.1}$$

is a pair of coordinates of one and only one point of the manifold in a neighborhood of the point P_0. Conversely: in a sufficiently small neighborhood of the point P_0, every point P has coordinates u, v satisfying inequality in (10.1).

The sphere and torus are two-dimensional manifolds without boundaries. A circular cylinder and a hyperbolic paraboloid are two-dimensional open manifolds (with boundaries at infinity). A sheet of a circular cone, the vertex excluded, forms a manifold which has one part of its boundary at infinity and another part at a finite distance (the vertex).

The volume contained inside a sphere constitutes an open three-dimensional manifold whose boundary is the surface of the sphere. When this volume includes the surfaces, it constitutes a three-dimensional manifold with a boundary, but the boundary forms part of the manifold. This is the reason that this type of manifold is called a *closed manifold*.

In the preceding examples, each manifold is defined as a set points situated in a pre-existent space. However, a manifold can be defined in *abstracto*. In the general case, an n-dimensional manifold is characterized by the existence of an analytic representation of a neighborhood of each point P_0 by means of a system of n coordinates u^i that may take arbitrary values in a neighborhood of the system of values $(u^i)_0$ representing the point P_0.

55. Analytic representation. There are infinitely many ways to choose coordinates that can analytically represent a region of a manifold. On passing from one of these coordinate systems to another, it is assumed that the new coordinates are continuous functions (functions of class C^0) of the original coordinates, and vice versa. *Topology* studies those properties of manifolds that are invariant under such transformations of coordinates.

In differential geometry new conditions are added: the new coordinates considered as functions of the original ones are not only continuous, but also admit continuous partial derivatives up to a certain order q (they are functions of class C^q). The range of properties that are invariant under such coordinate transformations is significantly more extensive than in topology.

Let us define, for example, a line giving coordinates of its points as functions of a parameter t. To say that these functions are differentiable with respect to the parameter t is to state a property of the line which is preserved under an arbitrary admissible coordinate transformation. We thus arrive to the notion of a *line element*. Analytically, a line element is defined by the n coordinates $u^1, \ldots, u^n$ and the mutual ratios of the differentials $du^1, \ldots, du^n$. Geometrically, a line element is defined by the collection of lines *tangent to each other* at a given point.

We arrive at the notion of a *plane element* by considering the set of line elements emanating from the same point and satisfying the same system of $n-2$ equations which are linear with respect to the differentials $du^1, \ldots, du^n$. Evidently, it is a property of this set of line elements that is preserved under any admissible coordinate transformation. The planar elements of dimension 3, 4, and higher are defined in a similar way.

If four line elements tangent to the same planar element emanate from a given point, then the cross-ratio (the anharmonic ratio) of these line elements is a number which is preserved under an arbitrary coordinate change. These considerations can be generalized in different ways.

Summarizing, we can say that the study of the properties of this nature is the geometry of the manifold from the point of view of the group of *continuous* and *differentiable* point transformations, whereas *topology* is the geometry of the manifold from the point of view of the group of simply *continuous* point transformations.

If we assume that new coordinates admit continuous partial derivatives of the two first orders with respect to original coordinates (they are functions of class C^2), and conversely, then the range of geometric notions extends significantly. Then we can study lines tangent to one another with second-order tangency, etc.

56. Riemannian manifolds. Regular metric. A *Riemannian manifold* (or a *Riemannian space*) is a manifold to which a *metric* is attached. This means that in each region of the manifold, analytically represented by means of a coordinate system u^i, a quadratic differential form is given,

$$ds^2 = g_{ij}du^i du^j.$$

We shall assume that the coefficients g_{ij} are functions of class C^2. Consequently, we shall only consider those coordinate transformations for which the new coordinates admit continuous partial derivatives of the two first orders with respect to the original coordinates (they are functions of class C^2), and conversely.

We shall say that *the metric is regular* in a given region of the manifold if at any point of this region, the quadratic form ds^2 is positive definite with respect to the differentials du^i.

Naturally, we shall assume that if the manifold consists of several regions admitting distinct analytic representations, then a concordance of metrics in the neighboring regions is possible. For example, it could be assumed that the analytic representation of each region could be extended by a small amount into the neighboring regions, and that two forms ds^2, thus obtained in the overlapping regions, could be transferred to one another by the coordinate transformation which takes one analytic representation to another.

Chapter 11

Locally Euclidean Riemannian Manifolds

57. Definition of a locally Euclidean manifold. A Riemannian manifold is said to be *locally Euclidean* if in each region of this manifold, defined analytically by a certain coordinate system u^i, the form ds^2 satisfies the conditions in (II) of No. **26** for a line element of the Euclidean space. On account of what has been discussed earlier, this means that the manifold, in a sufficiently small neighborhood of any of its point M_0, can be represented in a small domain of the Euclidean space with the same form ds^2.

We shall say that this representation constitutes a *development* of that region of the manifold in question in the Euclidean space. Conversely, we shall say that a domain of the Euclidean space is *applicable* on the corresponding small region of the manifold.

If the metric of the Riemannian manifold is everywhere regular, then we can develop the entire manifold piece by piece in the Euclidean space. However, one cannot be certain *a priori*

1°) That each point of the Euclidean space could be obtained in the development, and

2°) That a point of the Euclidean space obtained in the development of the manifold could not be obtained more than once.

58. Examples. Before going any further, consider a few simple examples.

1°. A *circular cylinder* in the ordinary space has the line element

$$ds^2 = du^2 + dv^2,$$

where u is the curvilinear abscissa ($0 \leq u \leq l$) taken along a cross-section normal to the cylinder axis, and v is the ordinate. This line element is Euclidean, but the manifold formed by the cylinder is not simply connected, and the development in the Euclidean plane produces a set of infinite strips of width l. Each point of the cylinder corresponds to an infinite number of points on the plane that are obtained, one from the other, by a translation with a fixed direction and whose length is an arbitrary multiple of the segment l.

We shall see that here the plane is covered entirely just once. We obtain an image of the manifold by taking in the plane an infinite strip in both directions, having a width l, and considering as identical two opposite points of the limiting parallels of the strip boundary when the straight line joining them is perpendicular to these parallels.

2°.   Another example is a *torus*. The position of a point on a torus is completely determined by two angles θ and φ (each of these angles varies from 0 to 2π). The first of these angles gives the rotation of the radius of a great circle, and the second gives a rotation of the radius of a meridian (see Fig. 4).

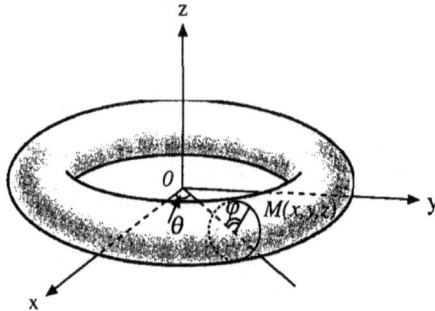

Fig. 4

Analytically the torus representation is

$$\begin{cases} x = (a - b\cos\varphi)\cos\theta, \\ y = (a - b\cos\varphi)\sin\theta, \\ x = b\sin\varphi. \end{cases} \tag{11.1}$$

By assigning a line element to the torus

$$ds^2 = a\,d\theta^2 + 2b\,d\theta d\varphi + c\,d\varphi^2 \tag{11.2}$$

with constant coefficients, we define a Euclidean metric on this manifold.[1] On the Euclidean plane, the variables θ and φ are the Cartesian coordinates. Thus, the torus develops in the Euclidean plane as a parallelogram

$$0 \le \theta < 2\pi; \quad 0 \le \varphi < 2\pi.$$

---

[1]Of course, it is understood that this line element has nothing in common with the line element induced by the ambient space in the analytic representation defined by the system of equations (11.1). Killing noted that this manifold (for $b = 0$) can be realized in the four-dimensional Euclidean space on the surface defined by the equations

$$x_1 = \sqrt{a}\cos\theta, \quad x_2 = \sqrt{a}\sin\theta, \quad x_3 = \sqrt{c}\cos\varphi, \quad x_4 = \sqrt{c}\sin\varphi.$$

Clifford has given another interpretation in three-dimensional elliptic space.

In order to define the development, one should restrict oneself to lines not crossing either the line $\theta = 0$ or the line $\varphi = 0$. If one removes these restrictions, the torus develops in the entire plane which is covered once and only once. However, the domain of the analytic representation of the manifold is a parallelogram whose opposite sides are not regarded as distinct.

3°. Our next example is an unbounded sheet of a *circular cone* whose line element is induced by the metric of the usual ambient space. As we know, this line element ds^2 is also Euclidean. Here it so happens that the *cone vertex is a singular point of the metric*, since a straight half-line emanating from the vertex and passing successively through all directions (on the cone) *describes an angle less than* 2π. If we wish to avoid investigating singular points of the metric, we must exclude the vertex from the manifold under consideration. Now the manifold will be *open* around the vertex (and around infinity), thus becoming from the point of view of topology, homeomorphic to a circular cylinder.

The development of this manifold on the Euclidean plane will now produce the entire plane (*with the exception of just one point*), but this plane is covered an infinite number of times (at least if the sine of the half-angle at the vertex is an irrational number).

4°. Finally, an arbitrary *developable surface* also has the Euclidean form ds^2, but its cuspidal edge is the locus of singular points of the metric. By taking only one of the sheets of the surface, we obtain a development covering only one region of the Euclidean plane. This development could cover this region several times or even an infinite number of times.

59. Riemannian manifold with an everywhere regular metric. The *distance* $[AB]$ between two points A and B in a Riemannian manifold with an everywhere regular metric is the lower bound of the lengths of rectifiable curves joining the points A and B. Obviously, for three arbitrarily given points A, B, and C, the following inequality holds:

$$[AC] \leq [AB] + [BC].$$

A *sphere* centered at A and with radius R is the set of points M satisfying the inequality

$$[AM] \leq R.$$

An infinite set of points of the space is said to be *bounded* if the distance from a fixed point A to the points of the set is bounded. This property is obviously independent of the choice of a fixed point A.

A point P of a Riemannian manifold is called a *limit point* of an infinite set (E) of points of this manifold if, in every sphere centered at P and of arbitrarily small radius r, there exists at least one point in the set (E) distinct from P. If this is the case, then there are an infinite number of such points.

60. Locally compact manifold.[2] A Riemannian manifold with an everywhere regular metric is said to be *locally compact* if any bounded infinite set of points of this manifold admits at least one limit point.

---

[2]In modern terminology, they are called *complete manifolds*. (Translator's note.)

Obviously, an unlimited sheet of a cone (*vertex excluded*), considered as a two-dimensional Riemannian manifold endowed with a metric induced by the (ordinary) Euclidean space in which the cone is embedded, does not enjoy the preceding property.

We shall say that a locally compact Riemannian manifold with an everywhere regular metric is *normal*. The cylinder, the torus (with the metric defined as above), and the Euclidean space itself are obviously normal. Two large classes of normal Riemannian manifolds exist:

1°. A normal Riemannian manifold is said to be *closed*, or *compact*, if every infinite set of points admits at least one limit point. In such a manifold, the distance $[AM]$ of a fixed point A to a variable point M is bounded.

2°. Nonclosed normal Riemannian manifolds are *open to infinity*. For example, the circular cylinder and Euclidean space itself are open to infinity. The expression "open to infinity" is self-explanatory: it states the existence of infinite sets of points moving indefinitely away from a given point A without having a limit point.

One can prove that if a normal Riemannian manifold with a Euclidean metric develops in the Euclidean space, then it covers the space once and only once.

Every simply connected normal Riemannian manifold with a Euclidean metric is identical to the Euclidean space.

61. The holonomy group. If a normal locally Euclidean Riemannian manifold is not simply connected, then in the development in the Euclidean space, a point M_0 gives several (or even an infinite number of) points

$$P_0, P_1, P_2, \ldots .$$

Each of these points is endowed with a Cartesian frame

$$(R_0), (R_1), (R_2), \ldots ,$$

corresponding to the same numerical values of the coefficients g_{ij} (they give the norms of the coordinate vectors and the cosines of the angles between them) at the point M_0. Thus all these frames are equal or symmetric.

If we had started the development by having the point M_0 correspond to the point P_1 with the frame (R_1), then the development of the Riemannian manifold would not have undergone any significant change. If a certain path (γ), going from M_0 to M, had produced in the first development a point P with a frame (R), then the new development would produce a point P' with a frame (R') situated with respect to the frame (R_1) in the same way as P and (R) were situated with respect to (R_0). This results in the displacement S_1 (whether or not followed by a symmetry), which in the Euclidean space superposes (R_0) with (R_1), varying between the frames

$$(R_0), (R_1), (R_2), \ldots$$

In fact, the frame (R_i) is obtained from (R_0) by the development of a certain closed contour (cycle) γ_i leaving the point M_0 and returning to the same point. The development of the same cycle leaving (R_1) will give a certain frame (R_j) situated with respect to (R_1) as the frame (R_i) was situated with respect to (R_0). The frame (R_i) could be obtained by first developing the cycle γ_1 and then the cycle γ_i.

The above considerations show that the displacements

$$S_1, S_2, S_3, \ldots$$

form a group G. In fact, we see that if we successively carry out the operations S_1 and S_i, then first the frame (R_0) coincides with (R_1), and second with (R_j). The resulting displacement is S_j:

$$S_1 S_i = S_j.$$

The group G thus obtained is called the *holonomy group* of the Riemannian manifold. To each frame (R_i), there is a corresponding single group operation. Namely, this is the operation superposing the frame (R_0) with the frame (R_i). To the frame (R_0), there corresponds an *identity operation*.

The operations of the holonomy group may be applied to any point in the Euclidean space, since the initial point P_0 of development is arbitrary. Two points corresponding to the same point of the Riemannian manifold are called *homologous*.

62. Discontinuity of the holonomy group of the locally Euclidean manifold. Let P be an arbitrary point in the Euclidean space. To this point, there is a certain corresponding point M in the Riemannian manifold. To the latter, we can assign a positive number r, such that every point P', distinct from P and situated at a distance from P less than r, is obtained from a point M' distinct from M. Consequently, *the distances from the point P of all points homologous of P are greater than r.*

This property proves that the holonomy group is discontinuous. It may be possible that for a discontinuous group, certain exceptional points exist that are themselves their own homologous points for a certain number of transformations of the group. However, here it is not the case.

Theorem 1. *The holonomy group of a locally Euclidean manifold is discontinuous, and all operations of this group, except the identity operation, do not leave any point of the manifold invariant.*

Chapter 12

Euclidean Space Tangent at a Point

63. Euclidean tangent metric. Consider a Riemannian manifold defined as a manifold endowed with an arbitrary line element

$$ds^2 = g_{ij}du^i du^j. \tag{12.1}$$

We shall assume that the right-hand side of (12.1) is a positive definite differential form with coefficients g_{ij} of class C^1.

The metric defined by a Euclidean line element

$$d\sigma^2 = \gamma_{ij}du^i du^j \tag{12.2}$$

constructed with variables u^i is called a *Euclidean tangent to the metric* (12.1) at the point $A(u_0^1, \ldots, u_0^n)$ if we have $\gamma_{ij} = g_{ij}$ for $u^i = u_0^i$.

Obviously, there exists an infinity of Euclidean metrics tangent to the metric (12.1) at a given point. In order to show this, it is sufficient, for example, to take

$$\gamma_{ij} = (g_{ij})_0.$$

On the other hand, the set of all these metrics does not depend on the choice of coordinates defining the Riemannian manifold analytically, since under the change of the variables u^i by certain other determined variables v^i, the new values at the point A of the coefficients of the line element only depend on their original values at A. Therefore, the equality $\gamma_{ij} = g_{ij}$ remains valid also for the new variables at the point A.

Rather than saying that we endow the manifold in question with a new (Euclidean) metric, we can say that we represent a Riemannian manifold in the Euclidean space in such a way that in this representation the line element of the Euclidean space becomes $d\sigma^2$. This Euclidean space is called the *Euclidean space tangent at the point A to the given Riemannian manifold*. This terminology is convenient since it provides a visualization.

We can say that there exists an infinity of Euclidean spaces tangent at the point A to a Riemannian manifold in the sense that the line element $d\sigma^2$ depends on an infinite number of arbitrary parameters. However, as we are going to consider in the sequel only geometric properties common to all these spaces, we may quite conveniently say, *the Euclidean space tangent at the point A to the Riemannian manifold.*

64. Tangent Euclidean space. The first geometric notion implied by consideration of the tangent Euclidean space is that of the distance between a point A and a point infinitesimally close to A. The distance is equal to $d\sigma$ or ds. It is the same notion on which the definition of a Riemannian manifold is based.

1°. *The angle between the two directions du^i and δu^i.* In the tangent Euclidean space, the value of the cosine of the angle between two vectors $\mathbf{X}$ and $\mathbf{Y}$ is obtained from the scalar product $\mathbf{X} \cdot \mathbf{Y}$ which coincides with the coefficient of 2λ in the decomposition of the square of the sum $\mathbf{X} + \lambda\mathbf{Y}$:

$$(\mathbf{X} + \lambda\mathbf{Y})^2 = \mathbf{X}^2 + 2\lambda\mathbf{X} \cdot \mathbf{Y} + \lambda^2\mathbf{Y}^2.$$

In the Riemannian manifold one must consider the product

$$g_{ij} \ (du^i + \lambda\delta u^i)(du^j + \lambda\delta u^j)$$
$$= g_{ij}(du^i du^j + \lambda^2 \delta u^i \delta u^j) + 2\lambda g_{ij} du^i \delta u^j.$$

Thus, the required cosine of the angle between two vectors $\mathbf{X}$ and $\mathbf{Y}$ is

$$\cos\varphi = \frac{g_{ij} du^i du^j}{ds \, \delta s}. \tag{12.3}$$

We can be sure in advance that *the right-hand side of this equation does not depend on the choice of a coordinate system,* since the angle between two directions in the tangent Euclidean space with the metric $d\sigma^2$ is determined by the same formula.

Likewise one may define the angle between a p-dimensional plane element and a q-dimensional plane element, etc., if they have a common vertex.

2°. We apply the notion of the angle between directions at a point of a Riemannian manifold in order to obtain the *rectangular coordinates in an orthogonal frame.*

As is known, a positive definite form

$$ds^2 = g_{ij} du^i du^j$$

can be always represented as a sum of squares:

$$ds^2 = (\omega^1)^2 + (\omega^2)^2 + \ldots + (\omega^n)^2, \tag{12.4}$$

where

$$\begin{cases} \omega^1 = a_1^1 du^1 + a_2^1 du^2 + \ldots + a_n^1 du^n, \\ \cdots\cdots\cdots\cdots\cdots\cdots\cdots\cdots\cdots\cdots\cdots\cdots\cdots \\ \omega^n = a_1^n du^1 + a_2^n du^2 + \ldots + a_n^n du^n. \end{cases} \qquad \det(a_i^j) \neq 0,$$

Consider now vectors $\mathbf{I}_i$ in the tangent Euclidean space, whose coordinates X^k satisfy the equations

$$
\text{for } \mathbf{I}_1: \quad
\begin{cases}
a_1^1 X^1 + \ldots + a_n^1 X^n = 1, \\
a_1^2 X^1 + \ldots + a_n^2 X^n = 0, \\
\cdots\cdots\cdots\cdots\cdots\cdots\cdots \\
a_1^n X^1 + \ldots + a_n^n X^n = 0;
\end{cases}
$$

$$
\cdots\cdots\cdots\cdots\cdots\cdots\cdots\cdots\cdots\cdots\cdots\cdots
$$

$$
\text{for } \mathbf{I}_n: \quad
\begin{cases}
a_1^1 X^1 + \ldots + a_n^1 X^n = 0, \\
a_1^2 X^1 + \ldots + a_n^2 X^n = 0, \\
\cdots\cdots\cdots\cdots\cdots\cdots\cdots \\
a_1^n X^1 + \ldots + a_n^n X^n = 1.
\end{cases}
$$

The expression in the numerator of formula (12.3) is a bilinear form associated with the quadratic form in (12.1). After reduction, the quadratic form in (12.1) takes the form (12.4). Then the bilinear form in the numerator of (12.3) becomes

$$
\omega^1(d)\,\omega^1(\delta) + \ldots + \omega^n(d)\,\omega^n(\delta), \tag{12.5}
$$

where

$$
\omega^i(d) = a_k^i du^k, \quad \omega^i(\delta) = a_k^i \delta u^k.
$$

Consider two vectors $\mathbf{I}_1$ and $\mathbf{I}_2$. For the first vector $\mathbf{I}_1$, we have

$$
\frac{\omega^1(d)}{dt} = 1, \quad \omega^2(d) = 0, \quad \ldots, \quad \omega^n(d) = 0,
$$

and for the second, we have

$$
\omega^i(\delta) = 0, \quad \frac{\omega^2(\delta)}{\delta t} = 1, \quad \ldots, \omega^2(\delta) = 0.
$$

Substituting these values into bilinear form (12.5), we see that this form vanishes. Thus, the vectors $\mathbf{I}_1$ and $\mathbf{I}_2$ are orthogonal (cf. No. **38**). The remaining identities

$$
\mathbf{I}_j \cdot \mathbf{I}_k = \delta_{jk}
$$

can be proved in a similar way. Thus, the frame constructed on the n vectors $\mathbf{I}_i$ in the tangent Euclidean space is orthonormal. The components of translations of this frame are the same forms ω^i that were used in the frame definition.

3°. At the point A of the Euclidean tangent space, one can define *vectors* (in an oblique frame, covariant and contravariant), *bivectors*, *multivectors*, and generally speaking, arbitrary *tensors*. All these can be extended to a Riemannian manifold. Moreover, we can introduce notions concerning the entire curve or the entire surface.

Firstly, since the infinitesimal distance between two infinitesimally close points is known, by means of addition (integration), we obtain the arc length of any curve,

$$s = \int \sqrt{\sum_i (\omega^i)^2}$$

This implies immediately (an idea due to Riemann) the notion of a *straight line*, or *geodesic* as giving an extreme for the distance. Here we merely indicate this possibility in order to return to it later.

Secondly, as is known, the elementary volume of the manifold is defined by the formula

$$d\tau = \omega^1 \wedge \omega^2 \wedge \ldots \wedge \omega^n.$$

65. The main notions of vector analysis. Given a vector field in a Riemannian manifold, it is not yet possible to define the derivative tensor, since, for this, one needs to be able to compare tensors at different points of the manifold. Nevertheless, we can define the gradient of a function and the *curl* of a vector field. For example, consider a scalar field. Let V be a function of a point,

$$V = V(u^1, \ldots, u^n).$$

The differential dV depends linearly on $du^1, du^2, \ldots, du^n$. Thus, this differential can be represented in the form

$$dV = V_1 \omega^1 + V_2 \omega^2 + \ldots + V_n \omega^n. \qquad (12.6)$$

The quantities V_i define the *gradient* of the function V.

Next, consider a field of vectors

$$(X_1, X_2, \ldots, X_n).$$

The tangent tensor form is

$$\widetilde{\omega} = X_1 \omega^1 + \ldots + X_n \omega^n.$$

Compute its exterior differential. Since the differential of X is linearly expressed in terms of ω^i, we have

$$d\widetilde{\omega} = X_{ij} \omega^i \wedge \omega^j, \qquad (12.7)$$

where X_{ij} is the skew-symmetric tensor called the *curl* of the vector field.

For example, for the expression

$$\widetilde{\omega} = X\, dx + Y\, dy + Z\, dz,$$

the exterior differential is

$$dX \wedge dx + dY \wedge dy + dZ \wedge dz$$

$$= \frac{\partial X}{\partial y} dy \wedge dx + \frac{\partial X}{\partial z} dz \wedge dx + \frac{\partial Y}{\partial x} dx \wedge dy$$

$$+ \frac{\partial Y}{\partial z} dz \wedge dy + \frac{\partial Z}{\partial x} dx \wedge dz + \frac{\partial Z}{\partial y} dy \wedge dz$$

$$= \left(\frac{\partial Z}{\partial y} - \frac{\partial Y}{\partial z} \right) dy \wedge dz + \left(\frac{\partial X}{\partial z} - \frac{\partial Z}{\partial x} \right) dz \wedge dx + \left(\frac{\partial Y}{\partial x} - \frac{\partial X}{\partial y} \right) dx \wedge dy,$$

i.e., we obtain the usual form of the curl in the Euclidean space.

We also derive the *divergence*. The flux of a vector field defines the tensor form

$$\tilde{\omega} = X_1 \omega^2 \wedge \omega^3 + X_2 \omega^3 \wedge \omega^1 + X_3 \omega^1 \wedge \omega^2,$$

where the exterior products $\omega^i \wedge \omega^j$ define certain areas.

The exterior differential has the form

$$d\tilde{\omega} = a \omega^1 \wedge \omega^2 \wedge \omega^3.$$

Here, a is the *divergence* of the vector field.

If we take the flux of the gradient,

$$V_1 \omega^2 \wedge \omega^3 + V_2 \omega^3 \wedge \omega^1 + V_3 \omega^1 \wedge \omega^2, \tag{12.8}$$

and apply exterior differentiation, then we obtain the second differential parameter

$$\Delta_2 V \omega^1 \wedge \omega^2 \wedge \omega^3.$$

EXAMPLE. Consider the line element

$$ds^2 = z^2 dx^2 + z^2 dy^2 + dz^2.$$

Here we have

$$\omega^1 = z\, dx, \quad \omega^2 = z\, dy, \quad \omega^3 = dz.$$

How can we find the second differential parameter?

In order to find the gradient, we differentiate V:

$$dV = \frac{\partial V}{\partial x} dx + \frac{\partial V}{\partial y} dy + \frac{\partial V}{\partial z} dz$$

$$= V_1 \omega^1 + V_2 \omega^2 + V_3 \omega^3 = V_1 z\, dx + V_2 z\, dy + V_3 dz.$$

Thus,

$$V_1 = \frac{1}{z} \frac{\partial V}{\partial x}, \quad V_2 = \frac{1}{z} \frac{\partial V}{\partial y}, \quad V_3 = \frac{\partial V}{\partial z}.$$

These expressions are the components of the gradient. The elementary vector flux is

$$V_1 \omega^2 \wedge \omega^3 + V_2 \omega^3 \wedge \omega^1 + V_3 \omega^1 \wedge \omega^2$$

$$= \frac{\partial V}{\partial x} dy \wedge dz + \frac{\partial V}{\partial y} dz \wedge dx + \left(z^2 \frac{\partial V}{\partial z}\right) dx \wedge dy.$$

Applying exterior differentiation, we find that

$$\Delta_2 V \omega^1 \wedge \omega^2 \wedge \omega^3 = \Delta_2 V z^2 dx \wedge dy \wedge dz$$

$$= \frac{\partial^2 V}{\partial x^2} dx \wedge dy \wedge dz + \frac{\partial^2 V}{\partial y^2} dy \wedge dz \wedge dx + \left(z^2 \frac{\partial V}{\partial z}\right) dz \wedge dx \wedge dy,$$

i.e., the second differential parameter is

$$\Delta_2 V = \frac{1}{z^2}\left(\frac{\partial^2 V}{\partial x^2} + \frac{\partial^2 V}{\partial y^2} + \left(z^2\frac{\partial V}{\partial z}\right)\right). \tag{12.9}$$

The function V is harmonic in this Riemannian manifold if $\Delta_2 V = 0$.

66. Three methods of introducing a connection. Although we were able to extend many notions from the tangent Euclidean space to a Riemannian manifold, there still remain many basic elementary notions that are missing: for example, we need to define the notion of an angle of rotation of a vector field when we move from one point of the manifold to another one.

In general, every geometric notion involving a *scalar* at any point can be generalized easily. The same can be said about a notion involving one vector or several vectors *provided that they have the same initial point.*

It seems as if the divergence of a vector field is an exception. However, in fact, *the elementary vector flux introduces a field at only one point.*

Summarizing, we see that until now a Riemannian manifold was for us a *collection of small pieces of the Euclidean space.* In some sense it was *amorphous,* since we have not yet connected these different pieces in a mutual orientation. Toward this end, we must define the rotation of a frame under a displacement of a point in the manifold. This can be done by different methods.

Riemann began with the postulate according to which a geodesic of the space, an extreme for the arc length, is not only the shortest line but also is the straightest line, i.e., it does not rotate a frame.

Another method is the same one that led us to the Euclidean tangent space: distinguishing an osculating Euclidean metric and extending its connection to a Riemannian manifold.

The third method is axiomatic: combining more or less natural requirements. This may be the most interesting method, since it admits a generalization to spaces with other fundamental groups.

Here we outline the second and the third methods.

67. Euclidean metric osculating at a point. A Euclidean metric osculating a given metric at a point $A(u_0^i)$ is defined by the line element

$$d\sigma^2 = \gamma_{ij} du^i du^j,$$

provided that the coefficients γ_{ij} and their first-order partial derivatives have the same numerical values at the point $A(u_0^i)$ as they have for the given line element.

The existence of osculating Euclidean metrics does not depend on the choice of coordinate system, since for a given change of variables, the new numerical values of the coefficients g_{ij} and their first-order partial derivatives are completely determined if the original values of the same quantities are known. Thus, rather than speak of an osculating Euclidean metric, we would rather say an *osculating Euclidean space*.

First, it is necessary to prove the existence of osculating Euclidean spaces at a given point A of Riemannian manifold.

From formulas (1.9), (1.10), and (5.4) for an oblique frame in the Euclidean space, we have

$$\begin{cases} d\mathbf{M} = \omega^i \mathbf{e}_i, \\[2mm] d\mathbf{e}_i = \omega_i^k \mathbf{e}_k, \quad \omega_i^k = \gamma_i{}^k{}_j \omega^j, \end{cases} \tag{12.10}$$

$$g_{ij} = \mathbf{e}_i \cdot \mathbf{e}_j, \quad dg_{ij} = \omega_i^k g_{kj} + \omega_j^k g_{ik}. \tag{12.11}$$

If we choose the so-called *natural frame*, i.e., if we set

$$\omega^i = du^i, \tag{12.12}$$

then

$$d\omega^i = 0,$$

and the first structure equation (II) of No. **26** implies that

$$du^j \wedge \omega_j^i = 0.$$

It follows from these equations that

$$\omega_j^i = \Gamma_j{}^i{}_k du^k, \quad \Gamma_j{}^i{}_k = \Gamma_k{}^i{}_j, \tag{12.13}$$

$$dg_{ij} = (\Gamma_j{}^i{}_h g_{kj} + \Gamma_j{}^i{}_h g_{ik}) du^k. \tag{12.14}$$

This implies immediately that

$$\frac{\partial g_{ij}}{\partial u^l} = \Gamma_i{}^k{}_l g_{kj} + \Gamma_j{}^k{}_l g_{ik}. \tag{12.15}$$

Thus, if at the point $u^i = u_0^i$, the quantities γ_{ij} coincide with $(g_{ij})_0$, and the coefficients $\Gamma_{j\ k}^{\ i}$ of the Euclidean frame at the initial point coincide with the corresponding values $(\Gamma_{j\ k}^{\ i})_0$ of the Riemannian metric for $u^i = u_0^i$, then not only the coefficients g_{ij} of the line element but also their partial derivatives $\frac{\partial g_{ij}}{\partial u^k}$ are respectively equal in both metrics.

This is not difficult to prove. Since from equations (12.10) and (12.13), it follows that

$$\frac{\partial \mathbf{M}}{\partial u^i} = \mathbf{e}_i, \quad \frac{\partial^2 \mathbf{M}}{\partial u^i \partial u^j} = \Gamma_{i\ j}^{\ k} \mathbf{e}_k, \tag{12.16}$$

we need only to select a parametrization of the Euclidean space in such a way that for $u^i = u_0^i$, equations (12.16) are satisfied.

At an arbitrary point O of the Euclidean space, construct an oblique Cartesian coordinate system $(\mathbf{e}_1, \dots, \mathbf{e}_n)$ such that for $u^i = u_0^i$, the equations

$$\mathbf{e}_i \cdot \mathbf{e}_j = (g_{ij})_0 \tag{12.17}$$

hold. With respect to this coordinate system, define the coordinates x^i of an arbitrary (moving) point of the Euclidean space by means of the equations

$$x^i = u^i - u_0^i + \frac{1}{2}\left(\Gamma_{r\ s}^{\ i}\right)_0 (u^r - u_0^r)(u^s - u_0^s), \quad i = 1, \dots, n. \tag{12.18}$$

The line element of the Euclidean space relative to this system of curvilinear coordinates enjoys the desired properties: the coincidence of the values of γ_{ij} and g_{ij} at the point $u^i = u_0^i$ follows from the construction of the initial frame (in particular, from formulas (12.17)), and the coincidence of the values $\Gamma_{j\ k}^{\ i}$ for the Euclidean space and the Riemannian manifold follows from formulas (12.16) by means of equations (12.18). Thus, at the point $u^i = u_0^i$, the numerical values of the functions $\gamma_{ij}, \frac{\partial \gamma_{ij}}{\partial u^k}$ and $g_{ij}, \frac{\partial g_{ij}}{\partial u^k}$ are equal.

Chapter 13

Osculating Euclidean Space

68. Absolute differentiation of vectors on a Riemannian manifold. All the geometric notions common to the different osculating Euclidean spaces at a point are obviously intrinsic (invariant) geometric properties of the Riemannian manifold. This is the case for all properties depending on numerical values at the point A of the coefficients g_{ij} and $\Gamma_i{}^k{}_j$.

Let us imagine that in the osculating Euclidean space, the point A and the frame (R_0) attached to this point are fixed. Each point M in the Riemannian manifold is represented by a point $\overline{M}$ whose coordinates, with respect to the frame (R_0), are well defined to within an order of infinitesimals *greater than the second*. The frame attached to the point $\overline{M}$ in the Euclidean space, with respect to its magnitude and orientation, is well defined to within some order of infinitesimals *greater than the first*. Moreover, this frame is equal to the frame attached to the point M in the Riemannian manifold. Every vector of the Riemannian manifold attached to M is represented with the same degree of approximation by an equal vector.

In general, if M and N are any two points in a neighborhood of the point A in the Riemannian manifold and if the points representing M and N in the osculating Euclidean space lie inside a sphere centered at A and of radius r, then the scalar product of two vectors representing vectors attached to M and N are well defined to within infinitesimals of order of radius r inclusively.

In a Riemannian manifold, a vector **X** attached to a point A and a vector **X′** attached to an infinitesimally close point A' are represented in the osculating Euclidean space by two vectors, whose difference is well defined to within an order of infinitesimals greater than the first. This leads us to extend to an arbitrary Riemannian manifold the notion of the *absolute differential* (or the *covariant differential for the natural frame*) of a vector, or more generally, of a tensor. We shall continue for this absolute differential the same notation D which we used earlier.

From equations (7.17), the absolute differential is

$$DX^i = dX^i + X^k \omega_k^i \tag{13.1}$$

for a contravariant vector (X^i), and

$$DX_i = dX_i - X_k \omega_i^k \qquad (13.2)$$

for a covariant vector (X_i). In an orthonormal frame, both definitions coincide, since then

$$X^k = X_k \quad \text{and} \quad \omega_i^{\ k} = \omega_{ik} = \omega^i_{\ k} = -\omega_k^{\ i}.$$

The minus sign appears because of the change of the order of the indices.

It is important to note that in the Euclidean geometry, the absolute differential is the true differential, while in the geometry of Riemannian manifolds, one cannot be sure that this is the case.

In particular, denoting by $\mathbf{e}_i$ (see No. 4) the coordinate vectors, we have

$$D\mathbf{e}_i = \omega_i^k \mathbf{e}_k.$$

DEFINITION. Two vectors attached to the two infinitesimally close points A and A' are said to be *equipollent*[1] if the absolute differential of the first vector (when this first vector is transported to the second vector) is equal to zero.

Thus, the conditions for equipollence are

$$dX^i + X^k \omega_k^i = 0, \qquad (13.3)$$

or

$$dX_i - X_k \omega_i^k = 0. \qquad (13.4)$$

69. Geodesics of a Riemannian manifold. It is not difficult now to define the *acceleration* of a moving point in the Riemannian manifold. If $\mathbf{v}$ is the velocity vector, then the acceleration vector is $\frac{D\mathbf{v}}{dt}$. For example, in order to write the contravariant components γ^i of the acceleration vector with coordinates (u^i) in the natural frame, note that from (12.12), (12.13), and (12.14), it follows that the velocity vector is

$$\mathbf{v} = \frac{D\mathbf{M}}{dt} = \frac{\omega^i}{dt} \mathbf{e}_i = \frac{du^i}{dt} \mathbf{e}_i, \quad v^i = \frac{du^i}{dt}.$$

Applying equations (13.1) to the components v^i and taking into account (12.13), we have

$$\frac{Dv^i}{dt} = \frac{d^2 u^i}{dt^2} + \frac{du^i}{dt} \omega_k^i = \frac{d^2 u^i}{dt^2} + \frac{du^k}{dt} \Gamma^i_{kh} \frac{du^h}{dt},$$

and hence

$$\gamma^i = \frac{d^2 u^i}{dt^2} + \Gamma^i_{kh} \frac{du^k}{dt} \frac{du^h}{dt}. \qquad (13.5)$$

A moving point whose acceleration is zero throughout has the velocity which is constantly equipollent to itself. The lines of a Riemannian manifold, whose

---

[1] In modern terminology, the term *parallel* is used. (*Translator's note.*)

tangent remains equipollent to itself throughout, are called the *geodesics* (the "straight lines") of the manifold. If a point moves along such a curve and has its velocity equal to 1, then the acceleration is equal to zero. It follows that the equations of geodesics in Riemannian manifolds have the form

$$\frac{d^2 u^i}{dt^2} + \Gamma^i_{kh} \frac{du^k}{dt} \frac{du^h}{dt} = 0. \tag{13.6}$$

These equations comprise Cauchy's system for the unknown functions u^i and the independent variable t. Under the hypotheses of Cauchy's theorem, with initial conditions

$$u^i = (u^i)_0, \quad \frac{du^k}{dt} = (v^k)_0 \text{ for } t = t_0, \quad (u^i)_0, (v^k)_0 \text{ are constants}, \tag{13.7}$$

such a system has one and only one solution. If we note that the first condition of (13.7) defines an initial point, and the second defines an initial velocity, we obtain the following result.

Theorem 1. *Through a point of a domain of regularity of a Riemannian manifold and in any direction, there passes one and only one geodesic.*

70. Generalization of the Frenet formulas. Curvature and torsion.
The theory of curvature of lines can be extended without modification from Euclidean space to any Riemannian manifold.

In fact, let M be a point on a curve, $\mathbf{T}$ be a unit tangent vector to the curve at the point M, and ds be the arc element of the curve. The vector $\frac{D\mathbf{T}}{ds}$ is normal to the vector $\mathbf{T}$. This can be seen immediately if one uses the Euclidean space osculating at the point M to the Riemannian manifold.

Suppose that

$$\frac{D\mathbf{T}}{ds} = \frac{1}{\rho} \mathbf{N},$$

where $\mathbf{N}$ is a unit vector (*principal normal*), and $\mathbf{B}$ is the unit vector (the *binormal*), which is orthogonal to $\mathbf{T}$ and $\mathbf{N}$ and forms a right trihedron with them. Suppose also that

$$\frac{D\mathbf{N}}{ds} = \alpha \mathbf{T} + \beta \mathbf{N} + \gamma \mathbf{B},$$

$$\frac{D\mathbf{B}}{ds} = \alpha' \mathbf{T} + \beta' \mathbf{N} + \gamma' \mathbf{B}.$$

The relations

$$\mathbf{T} \cdot \mathbf{N} = 0, \quad \mathbf{T} \cdot \mathbf{B} = 0, \quad \mathbf{N} \cdot \mathbf{B} = 0, \quad \mathbf{N}^2 = 1, \quad \mathbf{B}^2 = 1$$

on absolute differentiation, yield the following equations:

$$\frac{1}{\rho} + \alpha = 0, \quad \alpha' = 0, \quad \gamma + \beta' = 0, \quad \beta = 0, \quad \gamma' = 0.$$

Setting now

$$\gamma = -\beta' = \frac{1}{\tau},$$

we obtain the generalized Frenet formulas

$$\begin{cases} \dfrac{D\mathbf{T}}{ds} = \dfrac{1}{\rho}\mathbf{N}, \\[2mm] \dfrac{D\mathbf{N}}{ds} = -\dfrac{1}{\rho}\mathbf{T} + \dfrac{1}{\tau}\mathbf{B}, \\[2mm] \dfrac{D\mathbf{B}}{ds} = -\dfrac{1}{\tau}\mathbf{N}. \end{cases} \qquad (13.8)$$

The quantities $\dfrac{1}{\rho}$ and $\dfrac{1}{\tau}$ in these formulas are respectively the *curvature* and *torsion* of the curve. The "straight lines" (geodesics) of the Riemannian manifold are curves of zero curvature. As for the curves of zero torsion, they are characterized by the following properties:

1°) The osculating plane element at a point M of the curve is parallel to the osculating plane element at an infinitesimally close point M'.

2°) The vectors

$$\mathbf{T} + D\mathbf{T}, \quad \mathbf{N} + D\mathbf{N},$$

that are the unit vectors taken along the tangent and the principal normal to the curve at the point M' become

$$\mathbf{T} + \frac{ds}{\rho}\mathbf{N}, \quad \mathbf{N} - \frac{ds}{\rho}\mathbf{T} + \frac{ds}{\tau}\mathbf{B}.$$

These new vectors lie in the osculating plane element at the point M if and only if *the torsion* $\dfrac{1}{\tau}$ vanishes.

If we represent the Riemannian manifold in the osculating Euclidean space at the point M, then the image of the given curve (C) is a certain curve (Γ) *having the same curvature at the point M as* (C). Moreover, the vectors $\mathbf{T}, \mathbf{N}$, and $\mathbf{B}$ on the curve (Γ) are the images of the similar vectors on the curve (C). The following explanation makes this even more clear. *Whether we adapt the given metric or an osculating Euclidean metric at the point M, the curve (C) has the same tangent, principal vector, binormal, and the same curvature as* (Γ). *But* (C) *does not have the same torsion with the two metrics.* In particular, for an observer who would adapt an osculating Euclidean metric at M, every straight line (every geodesic) in the Riemannian manifold represents a point of inflection at M.

71. The theory of curvature of surfaces in a Riemannian manifold. The classical theory of curvature of surfaces can be extended to Riemannian manifolds quite easily.[2] The different curves on a surface (S) passing through a given point M of (S), have the same principal normal and even the same curvature in both metrics, the metric of the Riemannian manifold or the osculating Euclidean metric, adapted to M. It follows that the laws which govern the variation of the curvature of these curves, when their tangents at the point M are varied, are the same as for the Euclidean space.

All curves tangent to one another have the same *normal curvature*. This normal curvature is equal to $\frac{\cos V}{\rho}$, where V is the angle of the principal normal to the curve with the surface normal, and ρ is the radius of curvature of the curve. This result is *Meusnier's theorem*.

There exist two orthogonal directions tangent to the surface corresponding to the maximum and minimum normal curvature. These directions are called the *principal directions*. The normal curvatures corresponding to these directions are the *principal curvatures*.

The *curvature lines* are the lines which, at each of their points, are tangent to one of principal directions at that point. If θ is the angle formed by a curve with the curvature line of the first family, then we have for this curve

$$\frac{\cos V}{\rho} = \frac{\cos^2 \theta}{R_1} + \frac{\sin^2 \theta}{R_2}, \tag{13.9}$$

where $\frac{1}{R_1}$ and $\frac{1}{R_2}$ are the two principal curvatures.

The *asymptotic curves* are those whose normal curvature is zero. The osculating plane element of such curves is tangent to the surface. The straight line (the geodesic) of the Riemannian manifold tangent to an asymptotic line at a point M has a second-order tangency at this point with the surface.

The curvature lines can also be characterized by the following condition: for any given point M on the curve, the tangent to the curvature line at M, the surface normal at M, and the surface normal at an infinitesimally close point M' of this line, once transported by parallelism from M' to M, lie in the same plane element.

Denote by ν the unit vector of the surface normal, and by $\mathbf{T}_1$ and $\mathbf{T}_2$ the unit vectors of the tangents to the curvature lines. Then, for a displacement in the direction of the vector $\mathbf{T}_1$ with length ds_1, we have

$$D_1\nu = -\frac{ds_1}{R_1}\mathbf{T}_1.$$

Then, for a displacement in the direction of the vector $\mathbf{T}_2$ with length ds_2, we have

$$D_2\nu = -\frac{ds_2}{R_2}\mathbf{T}_2.$$

These formulas represent a generalization to the Riemannian manifolds of the classical *Olinde Rodriques equations*.

---

[2]The theory of curvature of surfaces in the Euclidean space is the subject of Chapter 23.

For a displacement with length ds in the direction making an angle θ with the principal direction $\mathbf{I}_1$, in general we have

$$\frac{D\nu}{ds} = -\frac{\cos\theta}{R_1}\mathbf{T}_1 - \frac{\sin\theta}{R_2}\mathbf{T}_2. \tag{13.10}$$

Finally, the total curvature $\frac{1}{R_1 R_2}$ of the surface at a point M could be defined by the method applied by Gauss to the Euclidean space. Take a surface element $d\sigma$ enclosing the point M and the sphere of center M and of radius 1 in the Euclidean space tangent at M to the Riemannian manifold. Transport by parallelism to the point M the surface normals at the different points of the element $d\sigma$, and consider a small region of the sphere defined by the intersection with the sphere of the normals so transported. The limit of the ratio $\frac{d\omega}{d\sigma}$ of the area of this portion of the sphere to the given area $d\sigma$ is equal to the total curvature $\frac{1}{R_1 R_2}$. We could the consider $d\omega$ as the solid angle of the cone at the vertex M obtained by transporting by parallelism at M the normals emanating from the different points of the element $d\sigma$.

72. Geodesic torsion. The Enneper theorem. The notion of the *geodesic torsion* of a curve on a surface can be extended to the most general Riemannian manifolds.

In fact, denote by $\mathbf{T}, \mathbf{N}$, and $\mathbf{B}$ the unit vectors along the tangent, principal normal, and binormal to the curve respectively, and keep the same notations $\mathbf{T}_1, \mathbf{T}_2$, and ν for the unit vectors along the tangents to the principal directions and the surface normal. Finally, denote by θ the angle formed by the tangent to the curve with the tangent to the first principal direction, and by V the angle of the principal normal to the curve and the surface normal. The direction cosines of the vectors $\mathbf{T}, \mathbf{N}$, and $\mathbf{B}$ with respect to the vectors $\mathbf{T}_1, \mathbf{T}_2$, and ν are given in Table 13.1:

	$\mathbf{T}_1$	$\mathbf{T}_2$	ν
$\mathbf{T}$	$\cos\theta$	$\sin\theta$	0
$\mathbf{N}$	$-\sin\theta\sin V$	$\cos\theta\sin V$	$\cos\theta$
$\mathbf{B}$	$\sin\theta\cos V$	$-\cos\theta\cos V$	$\sin\theta$

Table 13.1

Let us start with the generalized Frenet formulas

$$
\begin{cases}
\dfrac{D\mathbf{T}}{ds} = \dfrac{1}{\rho}\mathbf{N}, \\[2mm]
\dfrac{D\mathbf{N}}{ds} = -\dfrac{1}{\rho}\mathbf{T} + \dfrac{1}{\tau}\mathbf{B}, \\[2mm]
\dfrac{D\mathbf{B}}{ds} = -\dfrac{1}{\tau}\mathbf{N}
\end{cases}
\tag{13.11}
$$

and the equation

$$
\frac{D\nu}{ds} = -\frac{\cos\theta}{R_1}\mathbf{T}_1 - \frac{\sin\theta}{R_2}\mathbf{T}_2.
$$

Differentiating the relations

$$
\mathbf{T}\cdot\nu = 0, \quad \mathbf{N}\cdot\nu = \cos V, \quad \mathbf{B}\cdot\nu = \sin V,
$$

we obtain two distinct equations:

$$
\frac{\cos V}{\rho} = \frac{\cos^2\theta}{R_1} + \frac{\sin^2\theta}{R_2},
\tag{13.12}
$$

$$
\frac{dV}{ds} + \frac{1}{\tau} = \left(\frac{1}{R_1} - \frac{1}{R_2}\right)\sin\theta\cos\theta.
\tag{13.13}
$$

The left-hand side of equation (13.13) is a quantity called the *geodesic torsion* of the curve. The form of the right-hand side shows that all curves with the same tangent have the same geodesic torsion. The geodesic torsion vanishes for the curvature lines.

If we apply equation (13.13) to an asymptotic line which is not a straight line ($V = \frac{\pi}{2}$), then we obtain

$$
\frac{1}{\tau} = \left(\frac{1}{R_1} - \frac{1}{R_2}\right)\sin\theta\cos\theta.
$$

But for an asymptotic line, we have

$$
\frac{\cos^2\theta}{R_1} + \frac{\sin^2\theta}{R_2} = 0,
$$

and this implies that[3]

$$
\frac{1}{\tau} = \pm\sqrt{-\frac{1}{R_1 R_2}}.
$$

---

[3]In fact, we have $\frac{1}{\tau} = \frac{\cos\theta}{R_1\sin\theta}$ or $\frac{1}{\tau^2} = -\frac{1}{R_1 R_2}$. (*Translator's note.*)

This is the *Beltrami-Enneper theorem* for an arbitrary Riemannian manifold.

The asymptotic lines emanating from a point M of a surface have at that point torsions equal in absolute values but with opposite signs. The common absolute value of these torsions is equal to the square root of the total curvature of the surface taken with the sign changed.

In particular, the torsion of a doubly-asymptotic line is constantly zero.

It is known that, in Euclidean space, if two families of asymptotic lines coincide, then the asymptotic lines are straight lines. This theorem is no longer true in an arbitrary Riemannian manifold.[4] Nevertheless, by means of the Enneper theorem, we can affirm that they are curves with zero torsion.

73. Conjugate directions. The theory of *conjugate tangents* also generalizes to an arbitrary Riemannian manifold. If M and M' are two infinitesimally close points on a curve (C) traced on a surface (S), then the plane element tangent to the surface (S) at the point M', being transported by parallelism at the point M, has a certain direction, conjugate to the direction of (C) at M, in common with the plane element tangent to (S) at M. Two conjugate directions are harmonically conjugate with respect to the asymptotic directions. Finally, when we pass from M to M', the absolute differential of the unit vector of the surface normal is perpendicular to the conjugate direction of the curve. All these properties become trivial if we represent the Riemannian manifold in the osculating Euclidean space at the point M.

74. The Dupin theorem on a triply orthogonal system. We conclude with the *famous Dupin theorem*. As is known, a triply orthogonal system is a system formed by three one-parameter families of surfaces that intersect at right angles. The Dupin theorem states that the intersecting curves of each pair of surfaces from two different families is a curvature line for each of these surfaces.

The analytic proof that we are going to present can be applied just as well to Riemannian manifolds as to Euclidean spaces.

Since the cosines of the angles between the two intersecting coordinate curves is zero, the line element of our manifold is

$$ds^2 = g_{22}(du^1)^2 + g_{11}(du^2)^2 + g_{33}(du^3)^2.$$

For example, the condition, expressing the fact that the coordinate line (C_1) (u^1 a variable) is a curvature line for the surface (S_1) (u^2 a constant), is that the vector

$$\mathbf{e}_3 + D_1\mathbf{e}_3 du^1,$$

normal to (S_1) at the point M' infinitesimally close to the point M on the line (C_1), must belong to the same plane element as the vectors $\mathbf{e}_3$ and $\mathbf{e}_1$. In other

---

[4]É. Cartan, *Sur les courbes de torsion nulle et les surfaces développables dans les espaces de Riemann*, Comptes Rendus Acad. Sc. **184** (1927), 138–141. (JFM **53**, p. 696.)

words, the coefficient of e_2 in the expression of D_1e_3 is zero. But we have

$$De_3 = \omega_3^1 e_1 + \omega_3^2 e_2 + \omega_3^3 e_3,$$

$$D_1e_3 = \Gamma_{31}^1 e_1 + \Gamma_{31}^2 e_2 + \Gamma_{31}^3 e_3.$$

Thus, we need to prove that $\Gamma_{31}^2 = 0$. But taking equation (12.14) for $i = 1$, $j = 2$, $h = 3$ and applying equations $g_{23} = g_{31} = g_{12} = 0$, we come to the equations

$$\Gamma_{13}^2 g_{22} + \Gamma_{23}^1 g_{11} = 0,$$

$$\Gamma_{21}^3 g_{33} + \Gamma_{31}^2 g_{22} = 0,$$

$$\Gamma_{32}^1 g_{11} + \Gamma_{12}^3 g_{33} = 0,$$

where the last two equations are obtained from the first by cyclic permutation of the indices 1, 2, 3. By (12.13), the coefficients Γ_{ij}^k are symmetric with respect to the lower indices. Thus, adding the first two equations and subtracting the third, we obtain

$$2\Gamma_{31}^2 g_{22} = 0, \quad g_{22} \neq 0,$$

and as a result, we have

$$\Gamma_{31}^2 = 0.$$

Chapter 14

Euclidean Space of Conjugacy along a Line

75. Development of a Riemannian manifold in Euclidean space along a curve. The development of a Riemannian manifold in Euclidean space along a curve opens up new possibilities.

Suppose that coordinates of a varying point of a curve are expressed in terms of a parameter t which takes the value $t = 0$ at an initial point A_0. In a Euclidean space, take an origin O to which we attach a Cartesian frame R_0 determined in size and shape by the numerical values of the coefficients g_{ij} at the point A_0. With each point $M(t)$, we associate a point M' of the Euclidean space and a Cartesian frame (e'_i) attached to M'. We shall start with equations (I) of No. **26** and suppose that the frame (e'_i) is natural. Then we have

$$\begin{cases} d\mathbf{M}' = du^i \mathbf{e}'_i, \\ d\mathbf{e}'_i = \omega_i^j \mathbf{e}'_j \end{cases} \qquad (14.1)$$

with the independent variable t and the initial conditions: for $t = 0$, the point M' coincides with the point O, and the vector (e'_i) coincides with the vector (e_i) of the frame (R_0). As in No. **28**, we can prove that in the frame $(R) = (e'_i)$ attached to the point $M'(t)$, the scalar products $e'_i \cdot e'_j$ of its vectors are equal to the coefficients g_{ij}. Furthermore, if M and M_1 are arbitrary infinitesimally close points on a given curve, and M' and M'_1 are their images in the Euclidean space, then two equipollent vectors at the points M and M_1 have for their images two equipollent (simply parallel) vectors with at M' and M'_1.

Since a Cartesian frame (R) is attached to each point M', in fact, we obtain a development in Euclidean space not only of the given line but also of an entire infinitesimally small neighborhood of this line in the Riemannian manifold. In order to get the absolute geometric variation of a vector, whose initial point describes the arc of the given line in the Riemannian manifold, *it suffices to construct the ordinary geometric difference between the two vectors obtained in the preceding development.*

Thus, we have obtained a rigorous and precise realization of this absolute geometric variation. Now we also know precisely what is meant by "transport by the principle of equipollence[1] of a vector along a given path."

76. The constructed representation and the osculating Euclidean space. We prove now that the development obtained at each point of the curve enjoys the properties of an osculating Euclidean space. This means that we prove that along the curve, the constructed Euclidean line element with the variables u^i has the same numerical values of the coefficients g_{ij} and their first-order partial derivatives as those of the given line element.

Consider the operation of development in the Euclidean space in more detail. For simplicity, we take $n = 3$, and without loss of generality, assume that the line is defined by the equations $u^1 = 0$, $u^2 = 0$.

The development in the Euclidean space gives us the point M' and the vectors e_1', e_2', e_3' as functions of u^3. By means of equations (14.1), we obtain

$$
\begin{cases}
\dfrac{d\mathbf{M}'}{du^3} = \mathbf{e}_3', \\[2mm]
\dfrac{d\mathbf{e}_i'}{du^3} = \Gamma_{i\,3}^{\ k}\mathbf{e}_k'.
\end{cases}
\tag{14.2}
$$

Next we find, in terms of functions u^1, u^2, and u^3, a point P of the Euclidean space satisfying the following initial conditions: for $u^1 = 0$, $u^2 = 0$,

$$
\begin{cases}
\mathbf{P} = \mathbf{M}', \\[2mm]
\dfrac{\partial \mathbf{P}}{\partial u^1} = \mathbf{e}_1', \quad \dfrac{\partial \mathbf{P}}{\partial u^2} = \mathbf{e}_2', \\[2mm]
\dfrac{\partial^2 \mathbf{P}}{(\partial u^1)^2} = (\Gamma_{1\,1}^{\ k})^*\mathbf{e}_k', \quad \dfrac{\partial^2 \mathbf{P}}{\partial u^1 \partial u^2} = (\Gamma_{1\,2}^{\ k})^*\mathbf{e}_k', \quad \dfrac{\partial^2 \mathbf{P}}{(\partial u^2)^2} = (\Gamma_{2\,2}^{\ k})^*\mathbf{e}_k'.
\end{cases}
\tag{14.3}
$$

where $(\Gamma_{i\,j}^{\ k})^*$ are the values of these coefficients for $u^1 = 0$, $u^2 = 0$.

Of course, these conditions are compatible with each other. It suffices, for example, to take

$$
\mathbf{P} = \mathbf{M} + u^1\mathbf{e}_1' + u^2\mathbf{e}_2' + \frac{1}{2}(u^1)^2(\Gamma_{1\,1}^{\ k})^*\mathbf{e}_k'
$$

$$
+ u^1 u^2 (\Gamma_{1\,2}^{\ k})^*\mathbf{e}_k' + \frac{1}{2}(u^2)^2(\Gamma_{2\,2}^{\ k})^*\mathbf{e}_k'.
$$

In this formula, Euclidean space is related to a system of curvilinear coordinates u^1, u^2, u^3. For $u^1 = 0$, $u^2 = 0$, the natural frame attached to the point P (which

---

[1]This is "parallel transfer" or "parallel transport" in modern terminology. (*Translator's note.*)

reduces here to the point M') is determined by the vectors $\frac{\partial \mathbf{P}}{\partial u^i}$, which reduce here to the vectors $\mathbf{e}'_1, \mathbf{e}'_2, \mathbf{e}'_3$. The coefficients of the line element ds^2 in the Euclidean space are the coefficients of the given line element of the Riemannian manifold. As for the coefficients $\Gamma_{i\ j}^{\ k}$ of the Euclidean space, they are obtained if one compares the coefficients of $\mathbf{e}'_k$ in the expression $\frac{\partial^2 \mathbf{P}}{\partial u^i \partial u^j}$. If the indices i and j are both distinct from 3, then by applying (14.3), we find the coefficients $\Gamma_{i\ j}^{\ k}$ in the Riemannian manifold. If $i \neq 3$ and $j = 3$, then on account of (14.2) and for $u^1 = 0$, $u^2 = 0$, we find that

$$\left(\frac{\partial^2 \mathbf{P}}{\partial u^i \partial u^3} \right)_0 = \frac{d}{du^3} \left(\frac{\partial \mathbf{P}}{\partial u^i} \right)_0 = \frac{d\mathbf{e}'_i}{du^3} = (\Gamma_{i\ 3}^{\ k})^* \mathbf{e}'_k,$$

and the conclusion is the same. Finally, if $i = j = 3$, then

$$\left(\frac{\partial^2 \mathbf{P}}{(\partial u^3)^2} \right)_0 = \frac{d\mathbf{e}'_3}{du^3} = (\Gamma_{3\ 3}^{\ k})^* \mathbf{e}'_k,$$

and again the conclusion is the same.

We see that *the determination of a Euclidean space of conjugacy along a given curve does not require integration, once the development of the curve on the Euclidean space is carried out.*

This result has numerous important consequences. First of all, *an observer, who moves only along the given line and is content to restrict his measurements to an immediate neighborhood of this line, would not be able to see that he is outside a Euclidean space as long as he neglects infinitesimals of an order greater than the first.*

Another important consequence is as follows: if, in a Riemannian manifold, one considers an arc of a curve infinitesimally close to the curve (C), then *the length of this arc, measured by the metric of the Riemannian manifold or by the Euclidean metric of conjugacy along the curve (C), is the same up to the second-order infinitesimals.* [2]

In fact, at a point infinitesimally close to the curve (C), the coefficients of the given line element and the line element of the Euclidean space of conjugacy are equal up to second-order infinitesimals. Thus, *the representation of the Riemannian manifold upon the Euclidean space of conjugacy preserves the distances measured in the neighborhood of the given line up to the same order.*

77. Geodesics. Parallel surfaces. When one develops a curve on a Euclidean space, at each of its points, the developed curve has the same curvature and the same torsion as the given curve, since in development, the absolute differential of a vector, whose initial point describes the curve, becomes the ordinary geometric variation of this vector. The generalized Frenet formulas

---

[2] It is necessary to assume than the two arcs of the curves in question correspond, point by point, in a way that at two corresponding points, the directions of the tangent elements are infinitesimally close.

become, in the Euclidean space of conjugacy, the *usual* Frenet formulas:

$$
\begin{cases}
\dfrac{d\mathbf{T}}{ds} = \dfrac{1}{\rho}\,\mathbf{N}, \\[2mm]
\dfrac{d\mathbf{N}}{ds} = -\dfrac{1}{\rho}\,\mathbf{T} + \dfrac{1}{\tau}\,\mathbf{B}, \\[2mm]
\dfrac{d\mathbf{B}}{ds} = -\dfrac{1}{\tau}\,\mathbf{N}.
\end{cases}
$$

Thus, a curve with zero torsion is so developed along a plane curve, and a curve with zero curvature is developed along a straight line.

The straight lines (the geodesics) of a Riemannian manifold are therefore the curves whose development is along a straight line. This immediately implies that the straight lines (the geodesics) in our definition are equivalent to the straight lines in Riemann's definition.

In fact, if a curve (C) is developed along a straight line, and in a Riemannian manifold we take a curve (C') infinitesimally close to (C), which emanates from a given point A of (C), and ends at another given point B of (C), then the image, *a curve in the Euclidean space of conjugacy*, has the same length as (C), up to second-order infinitesimals. But in this Euclidean space, the line, which is the image of the curve (C'), results from the variation of a straight line segment. Thus, its length is *the same* as the length of the line segment up to second-order infinitesimals. Consequently, in the Riemannian manifold, the first variation of the length of a segment of a straight line (a geodesic) is identically zero, and the "straight line" (the geodesic) realizes the extremum of the distance conforming to the classical definition of Riemann.

This implies another *consequence*: if we consider a geodesic arc AB and an infinitesimally close geodesic arc $A'B'$, then as *in the Euclidean space*, up to second-order infinitesimals, we have

$$
\text{arc } A'B' - \text{arc } AB = -AA' \cos(\widehat{AB, AA'}) - BB' \cos(\widehat{BA, BB'}).
$$

In particular, taking a constant length R starting from the point A along the different geodesics emanating from A, we obtain a certain surface similar to the sphere as a region between the extremities. *The tangent plane element to this surface at a point M is normal to the geodesic emanating from A and ending at the point M.* The same applies if we consider a family of geodesics generating the surface from A only. Namely, if we take a constant length on each of these geodesics, then the line element tangent at any point M in the region between extremities is orthogonal to the geodesic emanating from A and ending at M.

One other consequence is the generalization of the property of *parallel surfaces*. If we construct normal geodesics at different points of a surface and take a constant length on each of them, then the region of the points so obtained is a surface which, at each of its points, is normal to the corresponding geodesic. It follows that if a two-parameter family of geodesics is normal to a surface, then it is normal to an infinity of surfaces.

78. Geodesics on a surface. In general, what we did above for a line is impossible to do for a surface. *In general, there is no Euclidean space of conjugacy along a surface.* The reason for this is very simple: in general, it is impossible to develop a surface in Euclidean space. Nevertheless, *if a development in Euclidean space is possible, then there exists a Euclidean space of conjugacy along a surface.*

Note the final application of the Euclidean space of conjugacy. In a three-dimensional Euclidean space, take a surface (S) and *a geodesic (C) of this surface*, i.e., a line realizing the minimum of distance *on a surface*. In the Euclidean space of conjugacy along the curve (C), the line (C) has a stationary length with respect to all lines on (S) having their ends at two given points of (C). Thus, by a classical property of geodesics on a surface of the Euclidean space, the osculating plane at any point of the line (C) is normal to the surface. *This property*, which is valid for the Euclidean space of conjugacy, *is also valid for a Riemannian manifold.*

C. CURVATURE AND TORSION
OF A MANIFOLD

Chapter 15

Space with a Euclidean Connection

79. Determination of forms ω_i^j for given forms ω^i. In No. **66** we posed the problem of finding a connection of a Riemannian manifold and outlined certain ways to solve this problem. First, we discovered that this problem is equivalent to the problem of absolute differentiation of vectors and depends on the choice of coefficients $\Gamma_{i\ k}^{\ j}$ (sometimes called the object of connection[1] or the forms ω_i^j).

We made broad use of the notion of the metric osculating at a point and obtained a series of geometric results.

Now we want to investigate to what degree we need the postulate on extending a connection of an osculating metric into a Riemannian manifold.

In order not to complicate the situation by the choice of frame, we assume that an orthonormal frame is assigned to every point of the Riemannian manifold.

We saw (No. **64**) that it is sufficient to reduce a positive definite quadratic form to a sum of squares (this is always possible),

$$ds^2 = (\omega^1)^2 + (\omega^2)^2 + \ldots + (\omega^n)^2, \tag{15.1}$$

in order that the forms ω^i define an orthonormal frame.

Will this determine the forms ω_i^j ?

We saw that in the Euclidean space, the forms ω^i and ω_i^j satisfy structure equations (II) of No. **26**,

$$d\omega^i = \omega^k \wedge \omega_k^i, \tag{II$_1$}$$

$$d\omega_i^j = \omega_i^k \wedge \omega_k^j, \tag{II$_2$}$$

and of course, the orthonormality condition

$$\omega_i^j = -\omega_j^i. \tag{15.2}$$

---

[1]They are now called the connection coefficients. (*Translator's note.*)

119

Conditions (15.2) followed from our choice an orthonormal frame. So, we need to investigate two equations (II_1) and (II_2). If both of them are satisfied, then the space is Euclidean. Equation (II_2) does not contain the forms ω^i at all, but these forms determine the line element (15.1). Thus, it is natural to search for the forms ω_i^j satisfying the system of equations (II_1) and (15.2).

Theorem 1. *If the forms ω^i are given, then it is possible to find the $\frac{n(n-1)}{2}$ forms ω_i^j satisfying the system of equations (II_1) and (15.2), and such a solution of this system is unique.*

Proof. We assume that only one frame is associated with each point of the space. Hence the forms ω_i^j are linear combinations of the forms ω^i:

$$\omega_i^j = \gamma_i{}^j{}_k \omega^k. \tag{15.3}$$

We need to find the coefficients $\gamma_i{}^j{}_k$ in such a way that they satisfy the system of equations (II_1) and (15.2).

Exterior differentiation of the given forms ω^i gives us the exterior quadratic form with respect to the basis forms ω^k in the form [2]

$$d\omega^i = c^i_{[jk]}\omega^j \wedge \omega^k, \quad c^i_{[jk]} = c^i_{jk} - c^i_{kj}. \tag{15.4}$$

Then equations (II_1) take the form

$$c^i_{[jk]}\omega^j \wedge \omega^k = \omega^k \wedge \omega_k^i. \tag{15.5}$$

Substituting expressions (15.3) into equations (15.5) and (15.2) and comparing the coefficients in independent products $\omega^j \wedge \omega^k$, we obtain

$$c^i_{[jk]} = \gamma^i_{jk} - \gamma^i_{kj}, \tag{15.6}$$

$$0 = \gamma^i_{jk} + \gamma^j_{ik}. \tag{15.7}$$

Equations (15.6) and (15.7) determine γ^i_{jk} uniquely.

In fact, twice making the cyclic permutation of the indices i, j, k in equations (15.6) and applying relations (15.7), we find that

$$\begin{cases} c^i_{[jk]} = \gamma^i_{jk} + \gamma^k_{ij}, \\ c^j_{[ki]} = \gamma^j_{ki} + \gamma^i_{jk}, \\ c^k_{[ij]} = \gamma^k_{ij} + \gamma^j_{ki}. \end{cases}$$

---

[2] Here and in what follows, it is assumed that for the alternated coefficients $c^i_{[jk]}$ the summation is carried over all combinations (j, k) of indices j and k:

$$c^i_{[jk]}\omega^j \wedge \omega^k = \sum_{(j,k)} c^i_{jk}\omega^j \wedge \omega^k.$$

Adding the first two equations and subtracting the third, we obtain

$$\gamma^i_{jk} = \frac{c^i_{[jk]} + c^j_{[ki]} - c^k_{[ij]}}{2}. \tag{15.8}$$

80. Condition of invariance of line element. Suppose now that the quadratic form ds^2 is decomposed into a sum of squares in the most general way, i.e., with $\frac{n(n-1)}{2}$ parameters

$$v^1, v^2, \ldots, v^{\frac{n(n-1)}{2}}$$

(this is the same as the number of independent forms ω^j_i satisfying equations (15.2)). The difficulty of the problem depends on the fact that although the forms ω^i are linear combinations of $du^1, \ldots, du^n$, the coefficients of these linear combinations can depend on u^i and v^α. Thus, after exterior differentiation the differentials dv^α can occur. By hypothesis, the line element

$$ds^2 = (\omega^1)^2 + (\omega^2)^2 + \ldots + (\omega^n)^2$$

does not depend on v^α. We now discuss how to express this condition.

It is always possible to choose new forms $\widetilde{\omega}^i_k$, which can also depend also on v^α and dv^α, in such a way that

$$d\omega^i = \omega^k \wedge \widetilde{\omega}^i_k, \tag{15.9}$$

or in bilinear form,

$$d\omega^i(\delta) - \delta\omega^i(d) = \omega^k(d)\widetilde{\omega}^i_k(\delta) - \omega^k(\delta)\widetilde{\omega}^i_k(d). \tag{15.10}$$

Suppose that the symbol d of differentiation corresponds to differentiation with respect to the variables u^i, and the symbol δ corresponds to differentiation with respect to the variables v^α. Then, by hypothesis, we have

$$\omega^i(\delta) = 0.$$

Denote also

$$\widetilde{\omega}^i_k(\delta) = e^i_k, \quad \omega^i(d) = \omega^i.$$

Then equations (15.10) take the form

$$\delta\omega^i = -e^i_k \omega^k.$$

Differentiating ds^2 with respect to the parameter v^α, we find that

$$\delta(ds^2) = 2\sum_i \omega^i \delta\omega^i = -2e_{ki}\omega^i\omega^k.$$

By hypothesis, $\delta(ds^2) = 0$. Hence,

$$e_{ki}\omega^i\omega^k = 0.$$

As a result, for any choice of the forms ω^k, all coefficients of $(\omega^i)^2$ and $\omega^i\omega^j (i \neq j)$ vanish,

$$e_{ii} = 0 \ (i \text{ not summed}), \ \ e_{ij} + e_{ji} = 0.$$

This proves that for fixed u^i, we have

$$\widetilde{\omega}_i^j(\delta) + \widetilde{\omega}_j^i(\delta) = 0.$$

Therefore, for any change of secondary parameters v^α, this sum depends only on the differentials $du^1, du^2, \ldots, du^n$, i.e., it depends on the forms ω^i. Thus,

$$\widetilde{\omega}_i^j(\delta) + \widetilde{\omega}_j^i(\delta) = a_{ik}^j \omega^k, \ \ a_{ik}^j = a_{jk}^i. \tag{15.11}$$

Now we will show how to write the form ω_i^j as a sum of the form $\widetilde{\omega}_i^j$, which can depend on dv^α, and a linear combination of the forms ω^i:

$$\omega_i^j = \widetilde{\omega}_i^j + \lambda_{ik}^j \omega^k, \tag{15.12}$$

where the coefficients λ_{ik}^j are unknown functions.

Equations (15.2) give now

$$\widetilde{\omega}_i^j + \widetilde{\omega}_j^i + (\lambda_{ik}^j + \lambda_{jk}^i)\omega^k = 0,$$

or by means of (15.11),

$$(a_{jk}^i + \lambda_{ik}^j + \lambda_{jk}^i)\,\omega^k = 0, \tag{15.13}$$

i.e.,

$$a_{jk}^i + \lambda_{ik}^j + \lambda_{jk}^i = 0. \tag{15.14}$$

The difference between equations (II_1) and (15.9) is

$$\omega^k \wedge (\omega_k^i - \widetilde{\omega}_k^i) = 0,$$

or by means of (15.12),

$$\lambda_{kh}^i \, \omega^k \wedge \omega^h = 0,$$

i.e.,

$$\sum_{(k,h)} (\lambda_{kh}^i - \lambda_{hk}^i)\, \omega^k \wedge \omega^h = 0.$$

Since the forms $\omega^k \wedge \omega^h$ are linearly independent, we find that

$$\lambda_{kh}^i - \lambda_{hk}^i = 0. \tag{15.15}$$

Making in (15.14) the cyclic permutation of the indices i, j, k, we obtain

$$\begin{cases} a_{ik}^j + \lambda_{ik}^j + \lambda_{jk}^i = 0, \\ a_{ji}^k + \lambda_{ji}^k + \lambda_{ki}^j = 0, \\ a_{kj}^i + \lambda_{kj}^i + \lambda_{ij}^k = 0. \end{cases} \tag{15.16}$$

Subtracting the sum of the first two equations from the third and applying (15.15), we get

$$\lambda^j_{ik} = \frac{a^i_{kj} - a^j_{ik} - a^k_{ji}}{2}. \tag{15.17}$$

Here, from (15.11), the coefficients a^j_{ik} are symmetric with respect to the indices i and j—this also can be seen from equations (15.16). The solution (15.17) satisfies the system of equations (15.14) and (15.15). Thus, it is always possible to find a solution of equations (15.1) and (15.2). Theorem 1 is proved.   □

81. Axioms of equipollence of vectors. Can the same absolute differentiation $DX = (dX^i + X^k\omega^i_k)e_i$ correspond to two different decompositions of ds^2 into sums of squares? We saw that to any decomposition of ds^2 into sums of squares, there is a certain corresponding connection given by its components ω^j_i. These components, and only these components, satisfy the first group of structure equations (II_1). Since a connection defined by the osculating Euclidean space is also defined up to transformations of coordinate lines, and the components of this connection satisfy equations (II_1), the two connections coincide.

We proved above that this connection can be obtained as a solution of the system of equations (II_1) and (15.2). In order to generalize the notion of equipollence of vectors, we must use another approach.

Suppose that two infinitesimally close points M and N are given. Establish a correspondence of *equipollence* between vectors at M and N (Levi–Civita's parallelism) by the following five requirements:

1°. *If a vector X' is equipollent to a vector X, then a vector mX is equipollent to a vector mX', where m is an arbitrary factor.*

These and the remaining requirements can be written as the formulas:

1°. If $X \approx X'$, then $mX \approx mX'$.

2°. If $X \approx X', Y \approx Y'$, then $X + Y \approx X' + Y'$.

3°. If $X \approx X'$, then $X^2 \approx (X')^2$, i.e., *equipollent vectors have equal lengths.*

4°. *If $X \approx X'$, then the difference of the corresponding components of these vectors is infinitesimally small.*

5°. This requirement will be given later.

Conditions 1° and 2° show that the vectors X and X' are linearly related:

$$X^{i'} = \alpha^i_k X^k.$$

From condition 4°, it follows that the last equation can be written in the form

$$X^{i'} = X^i + \omega^i_k X^k, \tag{15.18}$$

where ω^i_k are infinitesimally small.

Now we formulate the fifth condition:

5°. *The components ω_k^i are linear with respect to the differentials of independent variables.*

Condition 3° shows the equality of sums of squares of components of the vectors $\mathbf{X}$ and $\mathbf{X}'$:

$$\sum_i (X^i)^2 = \sum_i (X^{i'})^2.$$

Substituting (15.18) into the last equation and using the distributive law, we get

$$\sum_i (X^i)^2 = \sum_i (X^i + \omega_k^i X^k)^2$$

$$= \sum_i (X^i)^2 + 2 \sum_{i,k} X^i X^k + \sum_i (\omega_k^i X^k)^2.$$

Cancelling out the first terms $\sum_i (X^i)^2$ and neglecting the last term (as infinitesimally small of second order), we get

$$\sum_{i,k} \omega_k^i X^i X^k = 0.$$

Summing in the last equation over the combinations (i,k) of i and k, we obtain

$$\sum_i \omega_i^i (X^i)^2 + \sum_{(i,k)} (\omega_k^i + \omega_i^k) X^i X^k = 0.$$

Since the vector X^i is arbitrary, we have

$$\omega_i^i = 0 \ (i \text{ not summed}), \quad \omega_k^i + \omega_i^k = 0. \tag{15.19}$$

It follows that if

$$\mathbf{X}' = \mathbf{X} + d\mathbf{X},$$

then

$$dX^i = \omega_k^i X^k,$$

or

$$X^{i'} = X^i + \omega_k^i X^k. \tag{15.20}$$

This formula has the same form as earlier, but now all ω_i^j are arbitrary skew-symmetric forms.

We come to the following conclusion: the correspondence of equipollence is defined by an *arbitrary system of skew-symmetric forms* ω_i^j.

82. Space with Euclidean connection. Suppose now that a certain equipollence is given, i.e., there is given a law of correspondence of vectors at a point M and at a neighboring point N. Then the scalar product of vectors is preserved,

$$\mathbf{X} \cdot \mathbf{Y} = \mathbf{X}' \cdot \mathbf{Y}'.$$

In fact, from conditions 1° and 2°, it follows that

$$(\mathbf{X} + \lambda\mathbf{Y})^2 = \mathbf{X}^2 + \lambda^2\mathbf{Y}^2 + 2\lambda\mathbf{X}\cdot\mathbf{Y}.$$

But from condition 1°,

$$\mathbf{X}^2 = (\mathbf{X}')^2, \quad \mathbf{Y}^2 = (\mathbf{Y}')^2.$$

As a result, we have

$$\mathbf{X}\cdot\mathbf{Y} = \mathbf{X}'\cdot\mathbf{Y}'.$$

Preservation of the scalar product (and preservation of norms of vectors) implies preservation of angles between corresponding vectors.

Note that we can define a position of the point N with respect to the frame associated with the point M by means of the components w^i of translation. If we fix a law of equipollence, then we can represent the points M and N and vectors emanating from M and N in the same Euclidean space. Two "equipollent" vectors become *really equipollent* (parallel) in this Euclidean space.

Thus, we have defined an *Euclidean connection* in a "generalized space", i.e., as in the Euclidean space, we have connected neighborhoods of two infinitesimally close points M and N. In general, for arbitrary points M and N, this is impossible. However, two points A and B that are not infinitesimally close can be connected by a *certain* path AB.

Then we can represent a neighborhood of a point M_1, which is infinitesimally close to the point A on the line AB. Next we can move from M_1 to M_2, and so on, step-by-step by the same method. (Note that the correspondence obtained depends on the choice of a path AB.)

How is the passage to the limit realized effectively?

83. Euclidean space of conjugacy. Suppose that a point μ, representing a variable point M of the arc AB, moves along an arc $\alpha\beta$ representing the arc AB. Then for a frame at the point μ, we have

$$\begin{cases} d\mu = w^1\mathbf{i}_1 + w^2\mathbf{i}_2 + \ldots + w^n\mathbf{i}_n, \\ d\mathbf{i}_i = w_i^k\mathbf{i}_k. \end{cases} \tag{15.21}$$

If the point M describes the arc AB in such a way that its coordinates are represented by continuously differentiable functions (of class C^1) of a parameter t, then

$$\begin{cases} w^i = p^i dt, \\ w_i^j = p_i^j dt, \end{cases} \tag{15.22}$$

where for a fixed path AB, the quantities p^i and p_i^j are completely determined functions of the independent variable t. Substituting these values into equations

(15.21), we obtain the Cauchy system

$$
\begin{cases}
\dfrac{d\mu}{dt} = p^k \, \mathbf{I}_k, \\[2mm]
\dfrac{di_i}{dt} = p_i^k \, i_k.
\end{cases}
\tag{15.23}
$$

If we integrate this system, then we come to the *Euclidean space of conjugacy* (see No. **75**).

Choosing a certain Euclidean connection, we can extend the tensor analysis to this connection defining the absolute differential of a vector by the formulas for the contravariant and covariant vectors (in an orthonormal frame these formulas coincide):

$$
\begin{cases}
DX^i = dX^i + X^k \omega_k^i, \\[2mm]
DX_i = dX_i - X_k \omega_i^k.
\end{cases}
\tag{15.24}
$$

84. Absolute exterior differential. The most important moment is the extension of the notion of absolute differentiation of differential forms to the space with Euclidean connection.

In a generalized manifold, one cannot add vectors with different initial points, if these points are not connected by certain paths.

Consider a point A inside the domain in question and take a family of paths connecting the point A with different points of the domain. Then we can consider integrals over surfaces, volumes, etc.

In fact, suppose that a vector $(\widetilde{\omega}^i)$ is associated with a point M of a surface S. This vector, after being transferred to the point A along a path MA by means of equipollence, becomes

$$
\xi^i = \alpha_k^i \widetilde{\omega}^k.
$$

The sum of these vectors on a p-dimensional surface S is represented by the integral

$$
\underbrace{\int \cdots \int}_{(p)} \xi^i = \underbrace{\int \cdots \int}_{(p)} \alpha_k^i \widetilde{\omega}^k.
$$

Applying the generalized Gauss theorem to this integral, we represent this integral in the form

$$
\underbrace{\int \cdots \int}_{(p+1)} (\alpha_k^i d\widetilde{\omega}^k + d\alpha_k^i \wedge \widetilde{\omega}^k).
\tag{15.25}
$$

Now we restrict ourselves to a domain that is sufficiently small in order to be able to consider only an element of the integral.

We will see that under these conditions the result does not depend on the choice of paths.

In fact, the general formula

$$\xi^i = X^k \alpha_k^i$$

shows that at the point A, where $\xi^i = X^i$, all α_k^i, except α_i^i (i not summed), vanish,

$$\alpha_k^i = 0, \quad k \neq i, \quad \alpha_i^i = 1 \ (i \text{ not summed}).$$

Thus the integrand in (15.15) reduces to the sum

$$D\widetilde{\omega}^i = d\widetilde{\omega}^i + d\alpha_k^i \wedge \widetilde{\omega}^k.$$

At a point M, which is infinitesimally close to the point A, we have

$$\xi^i = X^i + d\alpha_k^i X^h.$$

On the other hand,

$$X^i = \xi^i + \omega_k^i \xi^k.$$

Hence

$$d\alpha_k^i = \omega_k^i$$

and

$$D\widetilde{\omega}^i = d\widetilde{\omega}^i + \omega_k^i \wedge \widetilde{\omega}^k. \tag{15.26}$$

This generalizes the formula for the absolute exterior differential in the Euclidean space.

Similarly we have

$$D\widetilde{\omega}_i^j = d\widetilde{\omega}_i^j - \omega_i^k \wedge \widetilde{\omega}_j^k + \omega_k^j \wedge \widetilde{\omega}_i^k. \tag{15.27}$$

85. Torsion of the manifold. Take an infinitesimally close contour around the point A. Let M and M' be two infinitesimally close points on the contour. Associate with the point M an infinitesimally small vector $\overrightarrow{ds} = \overrightarrow{MM'}$. Its components are

$$\widetilde{\omega}^i = \omega^i.$$

Take the sum of all these vectors. Because the contour is infinitesimally small, we can replace this sum by one vector. In Euclidean space, this sum vanishes. Here we have

$$D\omega^i = d\omega^i + \omega_k^i \wedge \omega^k = d\omega^i - \omega^k \wedge \omega_k^i = \Omega^i. \tag{15.28}$$

This sum can be equal to zero—*this is an intrinsic property of the space*, since for all decompositions of ds^2 into a sum of squares (i.e., for any choice of ω^i), this sum is zero.

In order to satisfy this property, we need to choose ω_i^j such that

$$d\omega^i = \omega^k \wedge \omega_k^i.$$

We saw that this is always possible to achieve, and in only one way.

This is one of the possible Euclidean connections for ds^2. In fact, this was Riemann's choice of connection. This choice implies that the sum of vectors of a closed contour around a point is equal to zero.

However, this connection is not the only possible connection.

If one considers the general formula

$$d\omega^i = \omega^k \wedge \omega_k^i + \Omega^i,$$

where ω_k^i depend on secondary parameters v^α (defining a decomposition of ds^2 into a sum of squares), then the quadratic form Ω^i does not depend on these parameters, and hence it does not depend on the forms ω_k^i. The form Ω^i defines the *torsion* of the manifold.

86. Structure equations of a space with Euclidean connection. Let us investigate now how the structure equations (II$_2$),

$$d\omega_i^j = \omega_i^k \wedge \omega_k^j, \tag{II$_2$}$$

will be changed in the new connection.

Associate with a point M of a cycle Γ a vector $\mathbf{X}$ and consider the integral

$$\int_\Gamma D\mathbf{X}.$$

In the Euclidean space this integral vanishes.

For an arbitrary connection, the components of $D\mathbf{X}$ are

$$\widetilde{\omega}^i = dX^i + X^k \omega_k^i. \tag{15.29}$$

Applying formula (15.26) for the absolute exterior differential to form (15.29), we obtain

$$D\widetilde{\omega}^i = dX^k \wedge \omega_k^i + X^k d\omega_k^i + \omega_k^i \wedge (dX^k + X^h \omega_h^k)$$
$$= X^k(d\omega_k^i + \omega_h^i \wedge \omega_k^h).$$

Set

$$d\omega_k^i - \omega_k^h \wedge \omega_h^i = \Omega_k^i.$$

Then

$$\int D\mathbf{X}^i = \iint X^k \Omega_k^i.$$

The quadratic forms Ω_k^i are skew-symmetric. They define a rotation which gives to the vector X the variation $\int D\mathbf{X}$. Thus, in general, we have

$$d\omega^i = \omega^k \wedge \omega_k^i + \Omega^i, \tag{15.30}$$

$$d\omega_i^j = \omega_i^k \wedge \omega_k^j + \Omega_i^j. \tag{15.31}$$

87. Translation and rotation associated with a cycle. As our paths AM, we take the arcs AM of the cycle in question and develop this cycle on the line $\alpha\mu\alpha'$ of the Euclidean space (Fig. 5).

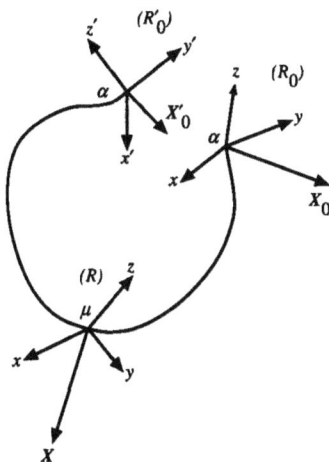

Fig. 5

A vector $\mathbf{X}$ attached to the point A is mapped into a vector $\mathbf{X}_0$ emanating from the point α and into a vector $\mathbf{X}_0'$ emanating from the point α'. When we move from M to M', we take $\mathbf{X}'$ at μ', $\mathbf{X}$ at μ, and compute

$$D\mathbf{X} = \mathbf{X}' - \mathbf{X},$$

i.e., in the Euclidean connection, we compute the ordinary difference $\mathbf{X}' - \mathbf{X}$, without moving to the point α. Thus,

$$\int D\mathbf{X},$$

is an ordinary integral for the path $\alpha\mu\alpha'$, i.e., it is the difference $\mathbf{X}_0' - \mathbf{X}_0$. We see that this result depends on a field value at the point A (dX^k disappears in the computation).

Next we move from the vector $\mathbf{X}_0$ to the vector $\mathbf{X}_0'$ by means of a simple infinitesimal rotation which takes the frame (R_0) attached to the point α into the frame (R_0') attached to the point α', i.e., it makes $\mathbf{X}_0$ parallel to $\mathbf{X}_0'$ (since the coordinates of $\mathbf{X}_0$ with respect to (R_0) are the same as the coordinates of $\mathbf{X}_0'$ with respect to (R_0')).

The quantities Ω_i^j are the components in (R_0') of an infinitesimally small rotation making $\mathbf{R}_0$ parallel to $\mathbf{R}_0'$. Similarly, the quantities Ω^i are the components in (R_0') of an infinitesimally small translation transferring α to α'.

Thus, the displacement taking (R_0) to (R_0') has components Ω^i and Ω_i^j.

This displacement vanishes in the Euclidean space. The *inverse* displacement (i.e., the displacement taking (R_0') to (R_0)) is called the displacement *associated with the cycle*.

Consider a field of equipollent vectors. In the Euclidean connection, the vector $\mathbf{Y}(\alpha')$ is equipollent in an ordinary sense (equal) to the vector $\mathbf{X}_0$. At the same time an observer of a Riemannian manifold considers the vector $\mathbf{X}_0$, which is not equipollent to the vector $\mathbf{Y}$, as the initial vector. The variation is $\mathbf{Y} - \mathbf{X}_0'$, i.e., the *rotation* Ω_i^j associated with the cycle.

Thus, the rotation associated with a cycle is the rotation undergone by a vector which is transferred along a cycle equipollently.

The associated rotation defines the *curvature*, and the associated translation defines the *torsion* at the point A.

88. The Bianchi identities. Consider structure equations (15.30) and (15.31) of the space with Euclidean connection.

Here

$$\Omega^i = R^i_{[jk]}\omega^j \wedge \omega^k, \quad \Omega_i^j = R^j_{i[kh]}\omega^k \wedge \omega^h,$$

where

$$R^i_{[jk]} = R^i_{jk} - R^i_{kj}, \quad R^j_{i[kh]} = R^j_{ikh} - R^j_{ihk}.$$

In order to obtain the Bianchi identities, we apply exterior differentiation to equations (15.30) and (15.31):

$$0 = d\omega^k \wedge \omega_k^i - \omega^k \wedge \omega_k^i + d\Omega^i,$$
$$0 = d\omega_i^k \wedge \omega_k^j - \omega_i^k \wedge d\omega_k^i + d\Omega_i^j.$$

In ordinary Euclidean space, these equations are identically satisfied. This implies that if we substitute the values of $d\omega^i$ and $d\omega_i^j$ from (15.30) and (15.31) into the last equations, all the terms, except those containing the forms Ω^i and Ω_i^j, must cancel out:

$$0 = \Omega^k \wedge \omega_k^i - \omega^k \wedge \Omega_k^i + d\Omega^i, \tag{15.32}$$

$$0 = \Omega_i^k \wedge \omega_k^j - \omega_i^k \wedge \Omega_k^j + d\Omega_i^j. \tag{15.33}$$

This is the first group of the Bianchi identities written for the space with Euclidean connection.

89. Theorem of preservation of curvature and torsion. We turn now to the interpretation of equations (15.32) and (15.33). Start from equations (15.33). They show that the absolute differential of the form Ω_i^j is equal to zero.

From equations (15.26), the absolute differential of the form $\widetilde{\omega}^i$ is

$$D\widetilde{\omega}^i = d\widetilde{\omega}^i + \omega_k^i \wedge \widetilde{\omega}^k.$$

Similarly, the absolute differential of the form $\widetilde{\omega}_i^j$ is

$$D\widetilde{\omega}_i^j = d\widetilde{\omega}_i^j - \omega_i^k \wedge \widetilde{\omega}_k^j + \omega_k^j \wedge \widetilde{\omega}_i^k.$$

Applying the last formula to the form $\widetilde{\Omega}_i^j$ and using (15.33), we find that

$$D\Omega_i^j = d\Omega_i^j - \omega_i^k \wedge \Omega_k^j + \omega_k^j \wedge \Omega_i^k = 0.$$

In fact, in the last formula, the first sum is the same as in (15.33). As to the second sum, we have $\omega_k^j \wedge \Omega_i^k = \Omega_i^k \wedge \omega_k^j$, since the form Ω_i^k is of even order, and this permutation of factors does not change the sign in the product.

Thus, Ω_i^j is the sum of bivectors (skew-symmetric tensors). Take a volume V bounded by a surface S. Consider bivectors Ω_i^j applied to the surface S. Their sum is equal to zero. This follows from equations (15.33) by means of the Gauss theorem:

$$\iint_S \Omega_i^j = \iiint_V d\Omega_i^j.$$

Now we turn to equations (15.32). The absolute differential of the form Ω^i is

$$D\Omega^i = d\Omega^i + \Omega^k \wedge \omega_k^i.$$

From equations (15.32), it follows that

$$D\Omega^i = \omega^k \wedge \Omega_k^i.$$

Consider, for example a three-dimensional space and an oriented element of a surface (i.e., an element with a direction of an infinitesimally small cycle). To every cycle, there are the corresponding translation Ω^i and the rotation Ω_i^j with components

$$\Omega^1, \Omega^2, \Omega^3; \ \Omega_2^3, \Omega_3^1, \Omega_1^2.$$

The rotation Ω_i^j can be represented by a vector $(\widetilde{\omega}^1, \widetilde{\omega}^2, \widetilde{\omega}^3)$, and the translation Ω^i can be represented as the moment of a pair.

Thus, on a surface, a double system of vectors is formed. Consider the sum of all vectors and all pairs. We need to connect a point A with all points M of a surface by certain paths.

If (X^1, X^2, X^3) is a vector, then we can transfer it to the point A by the formula

$$\xi^i = a_k^i X^k.$$

The components of a pair can be found from the matrix

$$\begin{pmatrix} x & y & z \\ a_k^1\widetilde{\omega}^k & a_k^2\widetilde{\omega}^k & a_k^3\widetilde{\omega}^k \end{pmatrix},$$

and they are

$$ya_k^3\widetilde{\omega}^k - za_k^2\widetilde{\omega}^k,$$
$$za_k^1\widetilde{\omega}^k - xa_k^3\widetilde{\omega}^k,$$
$$xa_k^2\widetilde{\omega}^k - ya_k^1\widetilde{\omega}^k,$$

where x, y, z are the coordinates of a point M with respect to the trihedron (R_0).

The sum of the vectors vanishes by means of equations (15.32). The sum of the pairs is

$$\iint_S [a_1^1\Omega^1 + a_2^1\Omega^2 + a_3^1\Omega^3 + y(a_1^3\Omega_2^3 + a_2^3\Omega_3^1 + a_3^3\Omega_1^2)$$
$$-z(a_1^2\Omega_2^3 + a_2^2\Omega_3^1 + a_3^2\Omega_1^2) + \ldots].$$

The integral along the entire surface reduces to the integral over the volume

$$\iiint_V [a_1^1 d\Omega^1 + da_1^1 \wedge \Omega^1 + \ldots + yd(a_1^3\Omega_2^3 + \ldots)$$
$$+ dy \wedge (a_1^3\Omega_3^2 + \ldots) - \ldots].$$

If a volume is very small, then by setting

$$a_1^1 = 1, \quad a_1^2 = a_1^3 = 0, \quad da_1^1 = 0, \quad da_i^k = \omega_i^k, \quad y = z = 0, \quad dy = \omega^2, \quad dz = \omega^3, \ldots,$$

we obtain

$$\iiint_V (d\Omega^1 + \omega_k^1 \wedge \Omega^k + \omega^2 \wedge \Omega_1^2 - \omega^3 \wedge \Omega_3^1).$$

But by means of the first equation of (15.32), we have

$$d\Omega^1 + \omega_k^1 \wedge \Omega^k = d\Omega^1 + \omega_k^1 \wedge \Omega^k - \omega^2 \wedge \Omega_2^1 - \omega^3 \wedge \Omega_3^1 = 0.$$

Thus, we have proved the following result.

Theorem 2. *In a Riemannian manifold with Euclidean connection, to a small volume there correspond both vectors and pairs, defining translations and rotations associated with a cycle on a surface. Geometrically, the sum of vectors and pairs is equal to zero.*

In mechanics one can find an interpretation in terms of the elastic forces acting on an area element.

Chapter 16

Riemannian Curvature of a Manifold

90. The Bianchi identities in a Riemannian manifold. The point of view of Riemann himself corresponds to a torsion-free connection,

$$\Omega^i = 0.$$

Now, the Bianchi identities take the form

$$\omega^k \wedge \Omega^i_k = 0, \tag{16.1}$$

$$\Omega^k_i \wedge \omega^j_k - \omega^k_i \wedge \Omega^j_k + d\Omega^j_i = 0. \tag{16.2}$$

The first of these identities show that there are no moments of pairs since the translation Ω^i is absent, and the second one shows that the sum of vectors vanishes (see No. **89**).

We decomposed ds^2 into a sum of squares and chose the unique forms ω^i_k in such a way that

$$d\omega^i = \omega^k \wedge \omega^i_k, \tag{16.3}$$

$$d\omega^j_i = \omega^k_i \wedge \omega^j_k + \Omega^j_i. \tag{16.4}$$

Is it possible to determine the forms ω^i_k independently of the choice of axes? This determination does not depend on the method of decomposition of ds^2 into a sum of squares. Take the most general decomposition of ds^2 into a sum of squares. The forms ω^j_i depend on the variables $u^1, u^2, \ldots, u^n$ and on parameters $v^1, v^2, \ldots, v^{\frac{n(n-1)}{2}}$ defining the method of decomposition. Obviously, Ω_{ij} can depend on both, ω^i and ω^j_i. We prove that Ω_{ij} depend only on ω^i. This follows from our Theorem 1 in No. **79**. We will prove this independently of this theorem.

From equations (15.31), we have

$$\Omega_i^j + \Omega_j^i = 0.$$

Denote the forms ω_{ij} by ω_α with Greek indices.

In general, the quadratic forms Ω_{ij} depend on both ω^i and ω_i^j. We can set

$$\Omega_{ki} = a_{ki[hl]}\, \omega^h \wedge \omega^l + b_{kil}^\alpha\, \omega^l \wedge \omega_\alpha + c_{ki}^{[\alpha\beta]}\, \omega_\alpha \wedge \omega_\beta,$$

where

$$a_{ki[hl]} = -a_{ki[lh]},$$
$$c_{ki}^{[\alpha\beta]} = -c_{ki}^{[\beta\alpha]}.$$

Substituting this expression for Ω_{ki} into equations (16.1), we find that

$$\omega^k \wedge \Omega_{ki} = a_{ki[hl]}\, \omega^k \wedge \omega^h \wedge \omega^l + b_{kil}^\alpha\, \omega^k \wedge \omega^l \wedge \omega_\alpha + c_{ki}^{[\alpha\beta]}\, \omega^k \wedge \omega_\alpha \wedge \omega_\beta.$$

Firstly, note that in the last sum, all terms contain different combinations of the basis forms $\omega^k, \omega_\alpha, \omega_\beta$, since the summation is carried over the combinations of α and β, and as a result, there are no similar terms. Thus,

$$c_{ki}^{[\alpha\beta]} = 0,$$

and the form Ω_{ki} does not contain terms with the products $\omega_\alpha \wedge \omega_\beta$.

Secondly, in the second sum for the triple of basis forms $\omega^k, \omega^l, \omega_\alpha$, two combinations

$$b_{kil}^\alpha\, \omega^k \wedge \omega^l \wedge \omega_\alpha + b_{lik}^\alpha\, \omega^l \wedge \omega^k \wedge \omega_\alpha$$

are possible. It follows that there is the symmetry in two lower indices:

$$b_{kil}^\alpha = b_{lik}^\alpha. \tag{16.5}$$

On the other hand, since the forms Ω_{ki} are skew-symmetric, we have

$$b_{kil}^\alpha = -b_{ikl}^\alpha. \tag{16.6}$$

Applying transformations (16.5) and (16.6) in turn, we obtain

$$b_{kil}^\alpha = b_{lik}^\alpha = -b_{ilk}^\alpha = -b_{kli}^\alpha = b_{lki}^\alpha = b_{ikl}^\alpha = -b_{kil}^\alpha.$$

It follows that

$$b_{kil}^\alpha = 0,$$

and the forms Ω_{ki} do not depend on the forms ω_α.

91. The Riemann–Christoffel tensor. From the previous section, we have

$$\Omega_i^j = R_i{}^j{}_{[kh]}\omega^k \wedge \omega^h. \tag{16.7}$$

The coefficients $R_{i\ [kh]}^{\ j}$ constitute a tensor, since if we contract these quantities with the vectors ω^k and ω^h, we obtain the tensorial differential form Ω_{ij}. This tensor is called the *Riemann–Christoffel tensor*.

Structure equations (16.3) and (16.4) can be written now in the form

$$d\omega^i = \omega^k \wedge \omega_k^i, \tag{16.8}$$

$$d\omega_i^j = \omega_i^k \wedge \omega_k^j + R_{i\ [kh]}^{\ j}\omega^k \wedge \omega^h. \tag{16.9}$$

If we substitute the values of the forms Ω_α from (16.7) into the first Bianchi identities (16.1), the latter give

$$R_{k\ hl}^{\ i}\omega^k \wedge \omega^h \wedge \omega^l = 0.$$

In order to collect similar terms, we cycle these equations with respect to the three lower indices. Equating the coefficients obtained to zero, we find that

$$R_{k\ hl}^{\ i} + R_{h\ lk}^{\ i} + R_{l\ kh}^{\ i} = 0,$$

or

$$(i) = R_{ikhl} + R_{ihlk} + R_{ilkh} = 0. \tag{16.10}$$

If we choose four arbitrary numbers i, k, h, l among $1, 2, \ldots, n$, then there will be a few dependencies among the components R_{ikhl} with the chosen indices. These indices are split into two pairs, i, k and h, l.

1°. *Under permutation of indices of one pair, the quantity R_{ikhl} changes its sign.* For the first pair i, k, this follows from the skew-symmetry of the forms Ω_i^k, and for the second pair h, l this follows from the accompanying permutation of factors ω^h, ω^l in the exterior product $\omega^h \wedge \omega^l$ and the invariance of the left-hand side Ω_k^i in (16.7).

2°. *Under permutation of pairs of indices, the quantity R_{ikhl} is not changed.* To prove this, consider equations (16.10) and equations obtained from these by the cyclic permutation of four indices i, k, h, l:

$$(i) = R_{ik,hl} + R_{ih,lk} + R_{il,kh} = 0,$$

$$(k) = R_{kh,li} + R_{kl,ih} + R_{ki,hl} = 0,$$

$$(h) = R_{hl,ik} + R_{hl,ik} + R_{hk,li} = 0,$$

$$(l) = R_{li,kh} + R_{li,kh} + R_{lh,ik} = 0.$$

Then

$$\frac{1}{2}[(i) + (k) - (h) - (l)] = R_{khli} - R_{likh} = 0. \tag{16.11}$$

This proves property 2°.

We turn now to the second group (16.2) of the Bianchi identities:

$$d\Omega_i^j + \Omega_i^k \wedge \omega_k^j - \omega_i^k \wedge \Omega_k^j = 0.$$

Substituting Ω_i^j from (16.7) into these identities, we obtain

$$dR_i^{\,j}{}_{kh} \wedge \omega^k \wedge \omega^h + R_i^{\,j}{}_{kh}\,\omega^i \wedge \omega_i^k \wedge \omega^h - R_i^{\,j}{}_{kh}\,\omega^k \wedge \omega^l \wedge \omega_i^h$$
$$+ R_i^{\,k}{}_{lh}\,\omega^l \wedge \omega^h \wedge \omega_k^j - R_k^{\,j}{}_{lh}\,\omega_i^k \wedge \omega^l \wedge \omega^h = 0.$$

Changing the summation indices, we have

$$(dR_i^{\,j}{}_{kh} - R_i^{\,j}{}_{lh}\,\omega_k^l - R_i^{\,j}{}_{kl}\,\omega_h^i + R_i^{\,l}{}_{kh}\,\omega_l^j - R_l^{\,j}{}_{kh}\,\omega_i^l) \wedge \omega^k \wedge \omega^h = 0.$$

Using the notation D for the absolute differential, we get

$$DR_i^{\,j}{}_{kh} \wedge \omega^k \wedge \omega^h = 0,$$

or

$$R_i^{\,j}{}_{kh|l}\omega^l \wedge \omega^k \wedge \omega^h = 0.$$

Cycling three indices l, k, h and equating the coefficients to zero, we obtain the second group of conditions for the covariant derivatives of the Riemann–Christoffel tensor:

$$R_i^{\,j}{}_{kh|l} + R_i^{\,j}{}_{hl|k} + R_i^{\,j}{}_{lk|h} = 0. \tag{16.12}$$

It follows that the number of all components of the Riemann–Christoffel tensor is:

$$\frac{n(n-1)}{2} \text{ of type } R_{ij,ij};$$

$$\frac{n(n-1)(n-2)}{2} \text{ of type } R_{ij,ik};$$

$$2\frac{n(n-1)(n-2)(n-3)}{4!} \text{ of type } R_{ij,kh};$$

altogether, we have

$$\frac{n^2(n^2-1)}{12}$$

different components.

It is always possible to find a Riemannian manifold corresponding to arbitrary values of these quantities.

92. Riemannian curvature. At a point A, consider a bivector defined by the coordinates $p^{ij} = -p^{ji}$. This is an oriented element of a surface. A certain rotation Ω_{ij} is associated with this element, and since

$$\omega^k \wedge \omega^h = p^{kh},$$

we have

$$q_{ij} = -R_{ij[kh]}p^{kh}.$$

Consider now the scalar product

$$p^{ij}q_{ij} = R_{ij,kh}p^{ij}p^{kh},$$

where $q_{ij} = -\Omega_{ij}$ are components of the associated rotation. The *Riemannian curvature* at the point A in the planar direction of the above bivector is defined by the formula

$$K = \frac{p^{ij}q_{ij}}{\sum\limits_{(i,j)}(p^{ij})^2} = -\frac{R_{ij,kh}p^{ij}p^{kh}}{\sum\limits_{(i,j)}(p^{ij})^2}. \tag{16.13}$$

93. The case $n = 2$. We have only one quadratic form, namely,

$$\Omega_1^2 = R_1{}^2{}_{12}\omega^1 \wedge \omega^2 = -K\omega^1 \wedge \omega^2. \tag{16.14}$$

This is the Riemannian curvature of the manifold.

We saw that a rotation is associated with a cycle. If $d\sigma$ is an area bounded by a cycle, then a vector $\mathbf{X}$, remaining equipollent to itself under traveling the cycle, will not return to the initial position after the cycle is complete. This vector takes new position $\mathbf{X}'$ forming an angle

$$K\,d\sigma$$

with the original position.

The structure equations are

$$\begin{cases} d\omega^1 = \omega^2 \wedge \omega_2^1, \\[2mm] d\omega^2 = \omega^1 \wedge \omega_1^2, \\[2mm] d\omega_1^2 = -K\,\omega^1 \wedge \omega^2. \end{cases}$$

In general, the Riemannian curvature is a function of a point. The simplest case is the space of constant curvature,

$$K = \text{const.}$$

An example of such a space is a sphere in an ordinary space. Its curvature $K = \frac{1}{R^2}$, where R is the radius of the sphere.

In fact, let R be a radius of a sphere, and O be its center. Associate with the point O trihedrons O, x_1, x_2, x_3 depending on three parameters.

If the point O is fixed, then a trihedron rotates with rotation components

$$\overline{\omega}_{12}, \ \overline{\omega}_{23}, \ \overline{\omega}_{31}.$$

If M is the common point of the third axis of the trihedron and the sphere, then

$$\overrightarrow{OM} = R\mathbf{I}_3$$

is the position vector of the point M of the sphere.

For the trihedrons in question, we have the following structure equations and the equations of infinitesimal displacement of axes:

$$
\begin{cases}
d\overline{\omega}_{23} = \overline{\omega}_{21} \wedge \overline{\omega}_{13}, & d\mathbf{I}_1 = \overline{\omega}_{12}\mathbf{I}_2 + \overline{\omega}_{13}\mathbf{I}_3, \\[2mm]
d\overline{\omega}_{31} = \overline{\omega}_{32} \wedge \overline{\omega}_{21}, & d\mathbf{I}_2 = \overline{\omega}_{21}\mathbf{I}_1 + \overline{\omega}_{23}\mathbf{I}_3, \\[2mm]
d\overline{\omega}_{12} = \overline{\omega}_{13} \wedge \overline{\omega}_{32}, & d\mathbf{I}_3 = \overline{\omega}_{31}\mathbf{I}_1 + \overline{\omega}_{32}\mathbf{I}_2.
\end{cases}
\qquad (16.15)
$$

If R is constant, then the displacement of the point M is equal to $d\mathbf{I}_3$ multiplied by R:

$$d\mathbf{M} = R\left(\overline{\omega}_3^1 \mathbf{I}_1 + \overline{\omega}_3^2 \mathbf{I}_2\right).$$

Thus, for the sphere, we have

$$\omega^1 = R\overline{\omega}_3^1, \ \ \omega^2 = R\overline{\omega}_3^2, \ \ \omega_1^2 = \overline{\omega}_1^2.$$

The structure equations are

$$d\omega^1 = R\,d\overline{\omega}_{31} = R\overline{\omega}_{32} \wedge \overline{\omega}_{21} = \omega^2 \wedge \overline{\omega}_{21},$$

$$d\omega^2 = \omega^1 \wedge \overline{\omega}_{12},$$

$$d\omega_1^2 = d\overline{\omega}_{12} = \overline{\omega}_{13} \wedge \overline{\omega}_{32} = -\frac{1}{R^2}\,\omega^1 \wedge \omega^2.$$

It follows that the curvature $K = \frac{1}{R^2}$.

This can be generalized to arbitrary dimension. All of these results can be obtained by considering a line element of a sphere. It has the form

$$ds^2 = R^2\,d\theta^2 + R^2 \sin^2\theta\,d\varphi^2,$$

where $\frac{\pi}{2} - \theta$ is the latitude, and φ is the longitude. Thus,

$$\omega^1 = R\,d\theta, \ \ \omega^2 = R\sin\theta\,d\varphi.$$

Applying exterior differentiation, we obtain

$$d\omega^1 = 0, \ \ d\omega^2 = R\cos\theta\,d\theta \wedge d\varphi.$$

Since

$$dw^1 = w^2 \wedge w_2^1, \quad dw^2 = w^1 \wedge w_1^2,$$

and hence

$$w_1^2 \wedge d\varphi = 0, \quad w_1^2 \wedge d\theta = \cos\theta \, d\varphi \wedge d\theta,$$

we find that

$$w_1^2 = \cos\theta \, d\varphi.$$

Further, we have

$$dw_1^2 = -\sin\theta \, d\theta \wedge d\varphi = -\frac{1}{R^2} w^1 \wedge w^2,$$

and thus the curvature $K = \frac{1}{R^2}$.

94. The case $n = 3$. Now equations (16.7) have the form

$$
\left\{
\begin{array}{l}
\Omega_{23} = R_{23,23} \, w^2 \wedge w^3 + R_{23,31} \, w^3 \wedge w^1 + R_{23,12} \, w^1 \wedge w^2, \\[2mm]
\Omega_{31} = R_{31,23} \, w^2 \wedge w^3 + R_{31,31} \, w^3 \wedge w^1 + R_{31,12} \, w^1 \wedge w^2, \\[2mm]
\Omega_{12} = R_{12,23} \, w^2 \wedge w^3 + R_{12,31} \, w^3 \wedge w^1 + R_{12,12} \, w^1 \wedge w^2.
\end{array}
\right.
$$

For brevity, denote $R_{23,23}$ by $-K_{11}$, $R_{31,31}$ by $-K_{12}$, and so on, replacing in each group of indices a pair of indices by a complementary index. Then we have

$$
\left\{
\begin{array}{l}
\Omega_{23} = -K_{11} \, w^2 \wedge w^3 - K_{12} \, w^3 \wedge w^1 - K_{13} \, w^1 \wedge w^2, \\[2mm]
\Omega_{31} = -K_{21} \, w^2 \wedge w^3 - K_{22} \, w^3 \wedge w^1 - K_{23} \, w^1 \wedge w^2, \\[2mm]
\Omega_{12} = -K_{31} \, w^2 \wedge w^3 - K_{32} \, w^3 \wedge w^1 - K_{33} \, w^1 \wedge w^2.
\end{array}
\right.
$$

The property $R_{ij,kh} = R_{kh,ij}$, yields the symmetry of K_{ij}:

$$K_{21} = K_{12}, \quad K_{31} = K_{13}, \quad K_{32} = K_{23}, \tag{16.16}$$

and the condition (16.12),

$$R_{i\ kh|l}^{\ j} + R_{i\ hl|k}^{\ j} + R_{i\ lk|h}^{\ j} = 0$$

takes the form

$$\sum_k K_{ik|k} = 0. \tag{16.17}$$

It should be pointed out that in equations (16.17), each of the first two indices represent a pair of complementary indices, and the third index (written on the right of the vertical line), the index of covariant differentiation, indicates the form w^k in which the covariant derivative stands for a coefficient.

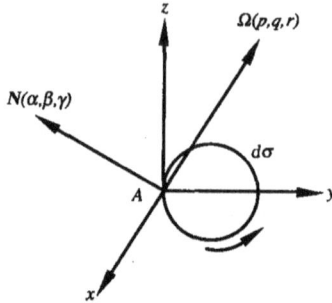

Fig. 6

95. Geometric theory of curvature of a three-dimensional Riemannian manifold. Suppose that a frame is connected with a point A and the element of surface $d\sigma$ passing through the point A (Fig. 6). Let $d\sigma$ be an oriented cycle, α, β, γ be the direction cosines of the normal, and $p\,d\sigma, q\,d\sigma, r\,d\sigma$ be the rotation associated with the cycle. We can always assume that $d\sigma$ is an infinitesimally small parallelogram with sides $\omega^i(d)$:

$$\begin{pmatrix} \omega^1(d) & \omega^2(d) & \omega^3(d) \\ \omega^1(\delta) & \omega^2(\delta) & \omega^3(\delta) \end{pmatrix}.$$

Then

$$\omega_2 \wedge \omega_3 = \alpha \, d\sigma.$$

Thus, if $\Omega(p, q, r)$ is the rotation associated with the cycle with components

$$p\,d\sigma, \quad q\,d\sigma, \quad r\,d\sigma,$$

then

$$\begin{cases} p\,d\sigma = K_{11}\,\alpha\,d\sigma + K_{12}\,\beta\,d\sigma + K_{13}\,\gamma\,d\sigma, \\ q\,d\sigma = K_{21}\,\alpha\,d\sigma + K_{22}\,\beta\,d\sigma + K_{23}\,\gamma\,d\sigma, \\ r\,d\sigma = K_{31}\,\alpha\,d\sigma + K_{32}\,\beta\,d\sigma + K_{33}\,\gamma\,d\sigma. \end{cases}$$

It follows that

$$\begin{cases} p = K_{11}\,\alpha + K_{12}\,\beta + K_{13}\,\gamma, \\ q = K_{21}\,\alpha + K_{22}\,\beta + K_{23}\,\gamma, \\ r = K_{31}\,\alpha + K_{32}\,\beta + K_{33}\,\gamma. \end{cases}$$

These formulas coincide with the formulas of the theory of elasticity which define the liquid pressure on a surface element.

Define the quadric-indicatrix (a surface of second order):

$$\Phi(x, y, z) = K_{11}x^2 + K_{22}y^2 + K_{33}z^2$$

$$+2K_{12}xy + 2K_{23}yz + 2K_{31}xz = 1.$$

The diametrical plane which is conjugate to the direction (α, β, γ) has the direction parameters

$$\frac{1}{2}\Phi'_\alpha, \ \frac{1}{2}\Phi'_\beta, \ \frac{1}{2}\Phi'_\gamma, \ \text{i.e., } p, q, r;$$

$$\Phi'_\alpha = \left.\frac{\partial \Phi}{\partial x}\right|_{x=\alpha, y=\beta, z=\gamma}, \quad \Phi'_\beta = \left.\frac{\partial \Phi}{\partial y}\right|_{x=\alpha, y=\beta, z=\gamma}, \quad \Phi'_\gamma = \left.\frac{\partial \Phi}{\partial z}\right|_{x=\alpha, y=\beta, z=\gamma}.$$

The equation of this plane is

$$x\Phi'_\alpha + y\Phi'_\beta + z\Phi'_\gamma = 0.$$

This implies the following result.

Theorem 1. *The vector representing the rotation associated with a surface element is normal to the diametrical plane which is conjugate to the direction of the surface normal. In order to find the magnitude of this rotation, it is sufficient to find its projection onto the direction (α, β, γ).*

Suppose that P is the common point of the quadric-indicatrix and the normal AP to a surface element. (A is a point at which the element is considered; this is the center of the quadric.)
We have

$$p\alpha + q\beta + r\gamma = \Phi(\alpha, \beta, \gamma) = \frac{1}{\overline{AP}^2}.$$

The *Riemannian curvature* at the point A in the direction of (α, β, γ) is equal to the projection of the rotation vector onto this direction: this is $\Phi(\alpha, \beta, \gamma)$ (Schur's definition).

The directions of the axes of the quadric-indicatrix are called the *principal directions* at the point A.

The rotation associated with an element normal to a principal direction takes place around this direction. If we take the coordinate axes as the quadric axes, then in the quadric equation, only squares remain. The coefficients in these squares are the *principal curvatures*. If the indicatrix is a sphere, then the principal curvatures are equal. In this case, the manifold is *isotropic* at a point.

96. Schur's theorem. *If a manifold is isotropic at each point, then the Riemannian curvature of the manifold is the same at all points.*

In fact, the vector Ω_{ij} is reduced to the form

$$-K\omega^i \wedge \omega^j.$$

The vanishing of its absolute exterior differential gives

$$dK \wedge \omega^i \wedge \omega^j = 0,$$

since

$$D\overline{\omega}_{ij} = d\overline{\omega}_{ij} + \omega_{ki} \wedge \overline{\omega}_{kj} + \omega_{kj} \wedge \overline{\omega}_{ik}.$$

All

$$D\overline{\omega}_{ij} = 0$$

by means of the second group of Bianchi's identities.

If

$$\overline{\omega}_{ij} = -K\omega^i \wedge \omega^j,$$

then

$$d\widetilde{\omega}_{ij} = -dK \wedge \omega^i \wedge \omega^j - K\,d\omega^i \wedge \omega^j + K\,d\omega^j \wedge \omega^i.$$

On the other hand, $D\overline{\omega}_{ij} = 0$ implies that

$$d\widetilde{\omega}_{ij} = -\omega_{ki} \wedge \overline{\omega}_{kj} - \omega_{kj} \wedge \overline{\omega}_{ik} = -K\,d\omega^i \wedge \omega^j + K\,d\omega^j \wedge \omega^i.$$

Thus,

$$dK \wedge \omega^i \wedge \omega^j = 0.$$

It follows from these equations that the differential dK depends linearly on ω^i and ω^j. But it does not depend on ω^i. Thus $dK = 0$, and $K = \text{const.}$

MECHANICAL INTERPRETATION. An ideal liquid in equilibrium, under action of internal forces only, is subject to a constant pressure at all points.

97. Example of a Riemannian space of constant curvature. Consider a rigid body rotating around a fixed point with velocity components p, q, r. If the ellipsoid of inertia is a sphere, then the kinetic energy has the form

$$2T = A\,(p^2 + q^2 + r^2),$$

and we have a Riemannian manifold with the line element

$$ds^2 = A\,(p^2 + q^2 + r^2)dt^2,$$

since

$$ds^2 = 2T\,dt^2.$$

Consider a coordinate trihedron invariantly connected with the rigid body in question. Suppose that $\widetilde{\omega}_{23}, \widetilde{\omega}_{31}, \widetilde{\omega}_{12}$ are the components of its rotation. We have

$$\widetilde{\omega}_{23} = p\,dt, \quad \widetilde{\omega}_{31} = q\,dt, \quad \widetilde{\omega}_{12} = r\,dt.$$

It follows that

$$ds^2 = A\,[(\widetilde{\omega}_{23})^2 + (\widetilde{\omega}_{31})^2 + (\widetilde{\omega}_{12})^2].$$

Since the rotation occurs in Euclidean space, we have

$$
\begin{cases}
d\omega_{23} = \omega_{21} \wedge \omega_{13}, \\
d\omega_{31} = \omega_{32} \wedge \omega_{21}, \\
d\omega_{12} = \omega_{13} \wedge \omega_{32}.
\end{cases}
$$

We can set

$$
\begin{cases}
\omega^1 = \sqrt{A}\,\omega_{23}, \\
\omega^2 = \sqrt{A}\,\omega_{31}, \\
\omega^3 = \sqrt{A}\,\omega_{12}.
\end{cases}
$$

This implies that

$$
\begin{cases}
d\omega^1 = -\dfrac{1}{\sqrt{A}}\,\omega^2 \wedge \omega^3, \\[2mm]
d\omega^2 = -\dfrac{1}{\sqrt{A}}\,\omega^3 \wedge \omega^1, \\[2mm]
d\omega^3 = -\dfrac{1}{\sqrt{A}}\,\omega^1 \wedge \omega^2.
\end{cases}
$$

Suppose that

$$
\omega_{23} = \lambda\omega_1, \quad \omega_{31} = \lambda\omega_2, \quad \omega_{12} = \lambda\omega_3.
$$

Then

$$
d\omega^1 = -\omega^2 \wedge \omega_1^2 - \omega_3^1 \wedge \omega^3 = -2\lambda\omega^2 \wedge \omega^3.
$$

Thus,

$$
\lambda = \frac{1}{2\sqrt{A}}
$$

and

$$
\omega_{23} = \frac{1}{2\sqrt{A}}\,\omega_1, \quad \omega_{31} = \frac{1}{2\sqrt{A}}\,\omega_2, \quad \omega_{12} = \frac{1}{2\sqrt{A}}\,\omega_3.
$$

It follows that

$$
\Omega_{23} = -d\omega_{23} - \omega_{21} \wedge \omega_{13},
$$

$$
= -\frac{1}{2A}\,\omega^2 \wedge \omega^3 + \frac{1}{4A}\,\omega^2 \wedge \omega^3 = -\frac{1}{4A}\,\omega^2 \wedge \omega^3.
$$

Finally, we have

$$
K_{ij} = 0, \quad i \neq j, \quad K_{ii} = \frac{1}{4A} \quad (i \text{ not summed}).
$$

We have a space of constant Riemannian curvature $\frac{1}{4\sqrt{A}}$.

98. Determination of the Riemann–Christoffel tensor for a Riemannian curvature given for all planar directions.

Theorem 2. *If a Riemannian curvature is given for all planar directions emanating from a point A, then all components $R_{ij,kh}$ of the Riemann-Christoffel tensor can be found.*

Proof. Note that this is not obvious: the quantities p^{ij} are not arbitrary, since they are the components of a bivector and are connected by certain known relations. Thus, one cannot claim *a priori* that knowing the curvature

$$R_{ij,kh}p^{ij}p^{kh}$$

implies knowing the components $R_{ij,kh}$.

In order to prove the theorem, we need to show that if the equation

$$R_{ij,kh}p^{ij}p^{kh} = R'_{ij,kh}p^{ij}p^{kh}$$

holds for all p^{ij}, then this implies that

$$R_{ij,kh} = R'_{ij,kh}.$$

In other words, if we set

$$S_{ij,kh} = R_{ij,kh} - R'_{ij,kh},$$

then it is sufficient to prove that the equation

$$S_{ij,kh}\, p^{ij}p^{kh} = 0 \qquad\qquad (16.18)$$

implies that

$$S_{ij,kh} = 0.$$

Introduce the vectors **X** and **Y** defining a bivector P with components

$$P^{ij} = X^i Y^j - X^j Y^i.$$

Then equation (16.18) takes the form

$$S_{ij,kh}\,(X^i Y^j - X^j Y^i)(X^k Y^h - X^h Y^k) = 0,$$

where all vectors X^i and Y^j are linearly independent. The last equation can be written as

$$S_{ij,kh}X^i Y^j X^k Y^h = 0.$$

In this sum (where the summation is carried out over all the indices i, j, k, and h), the term $(X^i)^2(Y^j)^2$ occurs only once. Thus, the coefficient of this term vanishes,

$$S_{ij,ij} = 0.$$

The term $(X^i)^2 Y^j Y^k$ occurs twice. Thus,

$$S_{ij,ih} + S_{ih,ij} = 0.$$

But using (16.11), we can easily prove that

$$S_{ij,ih} = S_{ih,ij}.$$

Hence,

$$S_{ij,ih} = S_{ih,ij} = 0.$$

Finally, the term $X^i X^k Y^j Y^h$ occurs four times and gives

$$S_{ij,kh} + S_{kj,ih} + S_{ih,kj} + S_{kh,ij} = 0,$$

or applying (16.11), we find that

$$2S_{ij,kh} + 2S_{ih,kj} = 0,$$

or

$$S_{ij,kh} = S_{ih,jk}.$$

Similarly,

$$S_{ik,hj} = S_{ij,kh}.$$

Thus, in the equations

$$S_{ij,kh} + S_{ik,hj} + S_{ih,jk} = 0,$$

which follows from (16.10), all terms are equal to one another. As a result, we have

$$S_{ij,kh} = 0.$$

99. Isotropic n-dimensional manifold. A manifold is called *isotropic* at a point A if the Riemannian curvature is the same in all directions, i.e., if

$$K = -\frac{R_{ij,kh}\, p^{ij} p^{kh}}{\displaystyle\sum_{(i,j)} (p^{ij})^2},$$

and K does not depend on p^{ij}, or if

$$R_{ij,kh}\, p^{ij} p^{kh} + K \sum_{(i,j)} (p^{ij})^2 = 0$$

for all bivectors p^{ij}.

This reduces to Theorem 1 if we set

$$R'_{ij,ij} = -K, \quad R'_{ij,kh} = 0.$$

As we proved, the components $R'_{ij,ij}$ must be equal to the corresponding components $R_{ij,ij}$.

Thus,

$$R_{ij,ij} = -K,$$

and all remaining components vanish,

$$R_{ij,kh} = 0.$$

Hence,

$$\Omega_{ik} = -K\omega^i \wedge \omega^j.$$

This is again Schur's theorem but now this theorem is extended to all values $n \geq 3$.

100. Curvature in two different two-dimensional planar directions.
Let q_{ij} and q'_{ij} be rotations associated with two surface elements p^{ij} and p'^{ij}. We have

$$p^{ij} q'_{ij} = R_{ij,kh}\, p'^{kh} p^{ij}.$$

The number

$$-\frac{R_{ij,kh}\, p'^{kh} p^{ij}}{\sqrt{\sum_{(i,j)} (p^{ij})^2}\, \sqrt{\sum_{(i,j)} (p'^{ij})^2}}$$

is called the *mixed curvature* in two directions. The curvature K is a special case, when two directions coincide.

101. Riemannian curvature in a direction of arbitrary dimension.
Until now we have considered only plane directions defined by bivectors.

Take now a three-dimensional element with which a system of trivectors

$$\Omega_{ijk} = \omega_i \wedge \Omega_{jk} + \omega_j \wedge \Omega_{ki} + \omega_k \wedge \Omega_{ij}$$

is associated. Consider a surface bounding this element, and attach a sliding (not free, as before) scalar bivector to each surface element.

The geometric sum of these bivectors gives a free trivector Ω_{ij}.

Consider the direction p^{ijk}, to which we attach a system of trivectors Ω_{ijk}, and consider the expression

$$K = -\frac{p^{ijk}\, \Omega_{ijk}}{\sqrt{\sum_{(i,j,k)} (p^{ijk})^2}} = -\frac{p^{ijk}\, R_{ik,hl}\, p_i^{hl}}{\sqrt{\sum_{(i,j,k)} (p^{ijk})^2}} = -\frac{R_{ik,hl}\, p^{ijk}\, p_i^{hl}}{\sqrt{\sum_{(i,j,k)} (p^{ijk})^2}}.$$

This is the corresponding Riemannian curvature.

The Riemannian curvature in a four-dimensional direction is defined in a similar way:

$$K = -\frac{R_{ik,kh}\, p^{ij,\rho\sigma}\, p_{\rho\sigma}^{kh}}{\sqrt{\sum_{(i,j,k)} (p^{ijk})^2}}.$$

102. Ricci tensor. Einstein's quadric. The important results are obtained for the dimension $n - 1$.

An $(n-1)$-dimensional direction is defined by means of an $(n-1)$-vector, or by means of an orthogonal vector. If α_i are the direction cosines of the normal, we can set

$$p_{234...n} = \alpha_1\, d\sigma, \quad p_{134...n} = -\alpha_2\, d\sigma, \ldots .$$

Then, in the expression of the curvature, the numerator is

$$K_{[i,j]}\sigma^i\sigma^j,$$

where

$$K_{12} = R_1{}^k{}_{2k}, \quad K_{11} = -R_{23,23} - R_{34,34} - \cdots .$$

Define the *Ricci tensor*

$$R_{ij} = R_i{}^k{}_{jk}.$$

We have

$$K_{ij} = R_{ij}, \quad i \neq j,$$

$$K_{11} = R_{11} - R^{ij}{}_{..,ij} = R_{11} - \frac{1}{2}\sum_i R_{ii}.$$

The tensor K_{ij} can be called the *Einstein tensor*.

We return to the notations of the case $n = 3$. We can define the *Einstein quadric*

$$K_{ij}X^iX^j = 1$$

and the *isotropy of the second kind* (the Einstein quadric is a sphere). There is also a theorem similar to Schur's theorem.

The notion of a space of constant curvature of the second kind is more general (only $\frac{n(n-1)}{2}$ coefficients).

Chapter 17

Spaces of Constant Curvature

103. Congruence of spaces of the same constant curvature. The structure equations of a space of constant curvature are

$$
\begin{cases}
d\omega^i = \omega^k \wedge \omega^i_k, \\
d\omega^j_i = \omega^k_i \wedge \omega^j_k - K\,\omega^i \wedge \omega^j.
\end{cases}
\tag{17.1}
$$

All coefficients in these equations are constant for an arbitrary choice of an orthonormal frame.

Theorem 1. *If there exists a space of a given constant curvature K, then such a space is unique.*

This means the following: if there are two spaces with curvature K and line elements ds^2 and $d\sigma^2$, then in some domain by means of an appropriate change of variables, we can reduce ds^2 to coincide with $d\sigma^2$. Take a neighborhood of a point M_0 of one of these spaces and an arbitrary point M'_0 of the second space. Then we can superpose the neighborhood of the point M_0 on the second space in such a way that the point M_0 coincides with the point M'_0 and a frame (R_0) associated with the point M_0 coincides with a frame (R'_0) associated with the point M'_0.

Proof. Suppose that $u^1, \ldots, u^N$, where $N = \frac{n(n+1)}{2}$, are parameters defining the most general frame at the point M_0, and $v^1, \ldots, v^N$ are similar parameters at the point M'_0. At the point M_0, we have equations (17.1), and at the point M'_0, we have the similar equations

$$
\begin{cases}
d\overline{\omega}^i = \overline{\omega}^k \wedge \overline{\omega}^i_k, \\
d\overline{\omega}^j_i = \overline{\omega}^k_i \wedge \overline{\omega}^j_k - K\,\overline{\omega}^i \wedge \overline{\omega}^j.
\end{cases}
\tag{17.2}
$$

Establish a correspondence between the points of two spaces by means of the equations

$$\begin{cases} \bar{\omega}^i(v, dv) = \omega^i(u, du), \\ \bar{\omega}^j_i(v, dv) = \omega^j_i(u, du), \end{cases} \tag{17.3}$$

where the v's must be considered as unknown functions of the u's. The system of equations (17.3) is completely integrable, since by equations (17.1) and (17.2), the equations

$$\begin{cases} d\bar{\omega}^i - d\omega^i = 0, \\ d\bar{\omega}^j_i - d\omega^j_i = 0 \end{cases} \tag{17.4}$$

are algebraic consequences of system (17.3).

In particular, system (17.4) admits solutions for which an initial system of values (u_0) corresponds to an arbitrary system of values (v_0), i.e., two arbitrary frames of the first and second family correspond to one another.

Since now

$$\begin{cases} v^1 = f^1(u^1, \dots, u^N), \\ \cdots\cdots\cdots\cdots\cdots\cdots\cdots \\ v^N = f^N(u^1, \dots, u^N), \end{cases}$$

we have a point correspondence between the two spaces.

In fact, the system of equations $\omega^i = 0$, $i = 1, \dots, n$, is completely integrable. Take $u^1, \dots, u^n$ as first integrals of this system. For constant values (u^i), we have $\omega^i = 0$, and the point M_0 of the first space is fixed. Thus, u^i are curvilinear coordinates of the first space. Take $v^1, \dots, v^n$ as those values of the first integrals of the system $\bar{\omega}^i = 0$ that correspond to the values (u^i) obtained in integrating of system (17.3). Since

$$\bar{\omega}^i = \omega^i, \tag{17.5}$$

v^i depend on $u^1, \dots, u^n$ only. To a point of the first space, there is a corresponding point of the second.

Finally, equations (17.5) imply the coincidence of the line elements,

$$ds^2 = \sum_i (\omega^i)^2 \text{ and } d\sigma^2 = \sum_i (\bar{\omega}^i)^2.$$

Corollary 2. *A space of constant curvature is applicable onto itself in an infinite number of ways: this space admits infinitely many motions (transformations preserving the distances). As with the Euclidean space, a space of constant curvature admits ∞^N displacements, where $N = \frac{n(n+1)}{2}$.*

REMARK. The spaces of constant curvature are the most general among the spaces admitting the group of motions. In fact, since the Riemannian curvature is defined by a line element ds^2, for displacements preserving ds^2, the Riemannian curvature is also preserved. In particular, a motion with a fixed arbitrary point A is possible. Thus, at this point A, the curvature in all directions is the same. Hence, the space is isotropic and has a constant curvature.

104. Existence of spaces of constant curvature. We pose this problem in another, more general form. The system of equations (17.1) is a particular case of a more general system of equations

$$d\omega^i = c^i_{[jk]}\omega^j \wedge \omega^k, \tag{17.6}$$

where

$$c^i_{[jk]} = c^i_{jk} - c^i_{kj}$$

are constants. The question is: does system (17.1) admit a solution ω^i ?

We obtain a necessary condition of existence of a solution, which is similar to the Bianchi identities, if we apply exterior differentiation to system (17.6):

$$0 = c^i_{[jk]}\left(d\omega^j \wedge \omega^k - \omega^j \wedge d\omega^k\right).$$

Changing in the second sum the summation indices $\binom{j\,k}{k\,j}$ and taking into account the fact that the coefficients $c^i_{[jk]} = c^i_{jk} - c^i_{kj}$ are skew-symmetric and the condition $\omega^j \wedge d\omega^k = d\omega^k \wedge \omega^j$ ($d\omega^k$ is of degree two), we obtain

$$0 = c^i_{[jk]}\, d\omega^j \wedge \omega^k.$$

Insert here the expression of $d\omega^i$ from formula (17.6). This gives

$$c^i_{[jk]}\, c^j_{[hl]}\, \omega^h \wedge \omega^j \wedge \omega^k = 0.$$

To collect similar terms, we make the circular substitution of indices h, l, k. This gives

$$\left(c^i_{[jk]}c^j_{[hl]} + c^i_{[jh]}c^j_{[lk]} + c^i_{[jl]}c^j_{[kh]}\right)\omega^h \wedge \omega^j \wedge \omega^k = 0,$$

or finally

$$c^i_{[jk]}c^j_{[hl]} + c^i_{[jh]}c^j_{[lk]} + c^i_{[jl]}c^j_{[kh]} = 0. \tag{17.7}$$

Equations (17.6) are called the *structure equations of the group of transformations*, and equations (17.7) are the *conditions for the structure constants* $c^i_{[jk]}$ *of the group*. The group transformations are obtained from the equations

$$\omega^i(u, du) = \omega^i(v, dv), \quad i = 1, 2, \ldots, r,$$

which, by equations (17.6), are completely integrable. Here (u^i) are the original coordinates of a point, and (v^i) are the coordinates of the transformed point.

We prove that conditions (17.7) are sufficient for the existence of r independent Pfaffian forms satisfying system (17.6). In group theory this corresponds to the third Sophus Lie theorem on existence of the group of transformations with given structure coefficients $c^i_{[jk]}$ satisfying conditions (17.7).

105. Proof of Schur. Suppose that we have a solution

$$\omega^i = A^i_k du^k,$$

where A^i_k are functions of independent variables u^k. Since the system of forms ω^i is linearly independent, the determinant of its coefficients is necessarily different from zero at the initial point:

$$\det(A^i_k) \neq 0 \quad \text{for} \quad u^k = 0. \tag{17.8}$$

Introduce a redundant number $r + 1$ variables instead of former r variables by setting

$$u^k = v^k t.$$

Then

$$\omega^i = A^i_k v^k \, dt + t A^i_k \, dv^k.$$

For $t = 0$, all coefficients in the differential dt are reduced to r functions

$$A^i_k(0) \, v^k,$$

that are independent by condition (17.8).

Define the quantities

$$a^i = A^i_k v^k, \tag{17.9}$$

which can be considered as new variables. So, now we have

$$\omega^i = a^i \, dt + \overline{\omega}^i(a, t, da), \tag{17.10}$$

where $\overline{\omega}^i$ are linear with respect to da^k and vanish if $t = 0$. We prove that for system (17.6), it is possible to construct solutions having the form of equations (17.10).

Taking exterior differentials of equations (17.10), we find that

$$d\omega^i = da^i \wedge dt + dt \wedge \frac{\partial \overline{\omega}^i}{\partial t} + \text{terms not depending on } dt. \tag{17.11}$$

In fact, by hypothesis, the form $\overline{\omega}^i$ can be written as

$$\overline{\omega}^i = M^i_k \, da^k.$$

Thus,

$$d\overline{\omega}^i = \left(\frac{\partial M^i_k}{\partial t} dt \right) \wedge da^k = \frac{\partial M^i_k}{\partial t} dt \wedge da^k$$

$$= dt \wedge \left(\frac{\partial M^i_k}{\partial t} da^k \right) = dt \wedge \frac{\partial \overline{\omega}^i}{\partial t}.$$

For a solution of equations (17.6), the terms containing dt on the right-hand side of (17.6) are

$$c^i_{[jh]} \, dt \wedge (a^j \, \overline{\omega}^h - a^h \, \overline{\omega}^j) = c^i_{jh} \, a^j \, dt \wedge \overline{\omega}^h, \qquad (17.12)$$

where on the right-hand side the summation is carried out with respect to the indices j and h separately.

Since t is the independent variable, the differential dt is also completely arbitrary. Thus if we substitute expressions (17.11) and (17.12) into equations (17.6), then the coefficients of dt on the left- and right-hand side of the equations obtained must coincide. The terms with dt are

$$da^i \wedge dt + dt \wedge \frac{\partial \overline{\omega}^i}{\partial t} = c^i_{jk} \, a^j \, dt \wedge \overline{\omega}^k.$$

It follows that

$$\frac{\partial \overline{\omega}^i}{\partial t} - da^i = c^i_{jk} \, a^j \, \overline{\omega}^k. \qquad (17.13)$$

Thus, the forms $\overline{\omega}^i$, considered as functions of t, satisfy a system of ordinary differential equations with constant coefficients. Therefore, there exists a unique solution of this system satisfying the initial conditions

$$\overline{\omega}^i = 0 \ \text{ for } \ t = 0.$$

Therefore, if there exists a solution, we can actually compute it.

106. The system is satisfied by the solution constructed. Suppose now that a solution $\overline{\omega}^i$ is obtained. We insert it into equations (17.10) and prove that system (17.6) is identically satisfied. In other words, if we set

$$d\omega^i = c^i_{[jk]} \, \omega^j \wedge \omega^k + \Omega^i, \qquad (17.14)$$

we must obtain

$$\Omega^i = 0.$$

In fact, we apply exterior differentiation to equations (17.14) and use the fact that the terms not containing Ω^i vanish by the imposed conditions which we assume are fulfilled. As a result, we obtain

$$d\Omega^i + c^i_{jk} \, \Omega^j \wedge \omega^k = 0. \qquad (17.15)$$

Introduce the symbol d for differentiation with respect to the variable t and the symbol δ_α for differentiation with respect to the variable a^α.

We prove that

$$\Omega^i(d, \delta_\alpha) = 0, \ \ \Omega^i(\delta_\alpha, \delta_\beta) = 0. \qquad (17.16)$$

To say that

$$\Omega^i(d, \delta_\alpha) = 0$$

is the same as to say that the terms containing dt are the same on the left-
and right-hand sides of equations (17.1). But comparison of these exact terms
produced equations (17.3) whose integration gave the forms $\overline{\omega}^i$.

In order to justify the second equation of (17.16), we apply equations (17.14)
to compute the bilinear form $\Omega^i(\delta_\alpha, \delta_\beta)$ for $t = 0$. We have

$$\Omega^i(\delta_\alpha, \delta_\beta) = \delta_\alpha \omega^i(\delta_\beta) - \delta_\beta \omega^i(\delta_\alpha)$$

$$-c^i_{[jk]} \left[\omega^j(\delta_\alpha) \omega^k(\delta_\beta) - \omega^j(\delta_\beta) \omega^k(\delta_\alpha) \right] = 0,$$

since for $t = 0$,

$$\omega^i(\delta_\alpha) = \omega^i(\delta_\beta) = 0.$$

Now the trilinear covariant with the symbols of differentiation $d, \delta_\alpha, \delta_\beta$ associated with equations (17.15) can be written as

$$d\,\Omega^i(\delta_\alpha, \delta_\beta) \ + \delta_\alpha\,\Omega^i(\delta_\beta, d) + \delta_\beta\,\Omega^i(d, \delta_\alpha) + c^i_{[jk]}\,[\Omega^j(d, \delta_\alpha)\,\omega^k(\delta_\beta)$$

$$+\Omega^j(\delta_\alpha, \delta_\beta)\,\omega^k(d) + \Omega^j(\delta_\beta, d)\,\omega^k(\delta_\alpha)] = 0.$$

Here, by the first (already proved) identity (17.16), all forms Ω^i for two symbols
of differentiation d and δ vanish.

We obtain

$$d\,\Omega^i(\delta_\alpha, \delta_\beta) + c^i_{jk}\,\Omega^j(\delta_\alpha, \delta_\beta)\,\omega^k(d) = 0.$$

If we substitute here the values of $\omega^k(d)$ taken from (17.10) and set

$$H^i = \Omega^i(\delta_\alpha, \delta_\beta),$$

then our equations become a system of ordinary differential equations:

$$\frac{dH^i}{dt} + c^i_j k\,H^j\,a^k = 0.$$

This is a linear homogeneous system. Thus, it admits the solution

$$H^i = 0.$$

But since we proved that for $t = 0$,

$$\Omega^i(\delta_\alpha, \delta_\beta) = 0,$$

it follows that the above solution is the only possible solution for any t.

Thus, we have proved the second identity of (17.16). This immediately
implies the vanishing of the exterior quadratic form Ω^i.

Now we need only to prove that all the forms ω^i we have found are linearly
independent.

Note that the variables a^i and t occur only as the products $a^i t$. In fact, equations (17.3), from which we found the forms ϖ^i, admit the substitution

$$t = kt', \quad a^i = \frac{\alpha^i}{k}, \quad k = \text{const.}$$

In order to integrate system (17.1), we integrate system (17.3) with the initial conditions

$$\varpi^i = 0 \text{ for } t = 0.$$

We get the forms w^i if we substitute $t = 1$ in the obtained solution ϖ^i: since

$$\varpi^i = A_k^i \, du^k, \quad \varpi^i = t A_k^i \, dv^k, \quad u^k = v^k t,$$

we have

$$\varpi^i = w^i \text{ for } t = 1.$$

Thus, we can make vanish those of a^i from (17.9) that correspond to the forms w_i^j with two indices.

In fact, from equations (17.10), we have

$$w^i = a^i \, dt + \varpi^i, \quad \varpi_i^j = w_i^j.$$

1°. For example, if $K = 0$ (the Euclidean space), the system of equations (17.1) becomes

$$dw^i = w^k \wedge w_k^i, \quad dw_i^j = w_i^k \wedge w_k^j, \tag{17.17}$$

and system (17.13) takes the form

$$\begin{cases} \dfrac{d\varpi^i}{dt} = da^i + a^k \, \varpi_k^i, \\[2mm] \dfrac{d\varpi_i^j}{dt} = 0 \end{cases} \tag{17.18}$$

In fact, compare equations (17.17) with system (17.6),

$$dw^i = c_{[jk]}^i \, w^j \wedge w^k,$$

and set

$$w_i^j = w^\alpha, \quad \alpha = n+1, \dots, N.$$

Note that

a) The coefficients $c_{[jk]}^i$ are equal to either 0 or 1.

b) For $i \leq n$ (the case of the first of equations (17.17)), the index j also satisfies the inequality $j \leq n$, and the index k satisfies the inequality $k \geq n+1$.

c) For $i > n$ (the case of the second of equations (17.17)), both indices j and k satisfy the inequality $j, k > n$, but for such large j, all $a^j = 0$.

Integrating system (17.18), we find that

$$\overline{\omega}^j_i = 0, \quad \overline{\omega}^i = t\,da^i,$$

and as a result, we have

$$\omega^i = da^i, \quad ds^2 = \sum_i (da^i)^2.$$

(This proves one more time that the forms ω^i are linearly independent.)

2°. If $K \neq 0$, then by (17.1), we obtain

$$\begin{cases} \dfrac{d\overline{\omega}^i}{dt} = da^i + a^k\,\overline{\omega}^i_k, \\[2mm] \dfrac{d\overline{\omega}^j_i}{dt} = -K\,(a^i\overline{\omega}^j - a^j\,\overline{\omega}^i). \end{cases} \qquad (17.19)$$

Multiplying the first equation of (17.19) by a_i and summing with respect to the index i, we find that

$$a_i\,\frac{d\overline{\omega}^i}{dt} = a_i\,da^i, \qquad (17.20)$$

since

$$a_i\,a^k\,\overline{\omega}^i_k = \sum_{(i,k)}(a_i\,a^k\,\overline{\omega}^i_k + a_k\,a^i\,\overline{\omega}^k_i) = \sum_{(i,k)} a_i\,a_k\,(\overline{\omega}^i_k + \overline{\omega}^k_i) = 0.$$

Integrating equation (17.20) with respect to t and taking into account that a^i does not depend on t, for the initial conditions $\omega^i = 0$ for $t = 0$, we obtain

$$a_i\,\overline{\omega}^i = t\,a_i\,da^i. \qquad (17.21)$$

Now, differentiating the first equation of (17.19) with respect to t, we get

$$\frac{d^2\overline{\omega}^i}{dt^2} = a^k\,\frac{d\overline{\omega}^i_k}{dt} = -K\sum_{(i,k)} a^k\,(a_k\,\overline{\omega}^i - a_i\,\overline{\omega}^k)$$

$$= -K\,a^k a_k\,\overline{\omega}^i + K\,a_i\,t\,a_k\,da^k.$$

The first transformation was made by means of the second equation of (17.19), and the second by equations (17.21).

Thus,

$$\frac{d^2\overline{\omega}^i}{dt^2} = -K\,a^k a_k\,\overline{\omega}^i + \varphi(t),$$

and we can integrate with the independent variable t and the unknown functions $\overline{\omega}^i$.

Chapter 18

Geometric Construction of a Space of Constant Curvature

107. Spaces of constant positive curvature. We can obtain a space of constant curvature by considering a hypersphere in an $(n+1)$-dimensional Euclidean space. We prove now that a hypersphere of radius R has the constant curvature $\frac{1}{R^2}$.

In an $(n+1)$-dimensional Euclidean space, consider a rectangular Cartesian coordinate system with an origin O and coordinates $x^1, \ldots, x^{n+1}$. The line element of this space is

$$ds^2 = (dx^1)^2 + (dx^2)^2 + \ldots + (dx^{n+1})^2.$$

The equation of the hypersphere with center at the origin and radius R is

$$(x^1)^2 + (x^2)^2 + \ldots + (x^{n+1})^2 = R^2.$$

To the fixed point O, we attach an orthonormal frame with unit basis vectors $\mathbf{I}_1, \ldots, \mathbf{I}_{n+1}$. Denote by M the common point of the hypersphere and the position vector of the direction $\mathbf{I}_{n+1}$. Then

$$\overrightarrow{OM} = R\,\mathbf{I}_{n+1}. \tag{18.1}$$

If the origin of our frame is fixed, the frame displacement is determined by the forms ω_α^β, $\alpha, \beta, \gamma = 1, \ldots, n+1$, satisfying the structure equations

$$d\omega_\alpha^\beta = \omega_\alpha^\gamma \wedge \omega_\gamma^\beta.$$

In what follows, the Greek letters take the values $\alpha, \beta, \gamma = 1, \ldots, n+1$, and the Latin letters take the values $i, j, k = 1, \ldots, n$. The equations of infinitesimal displacement of the point $\mathbf{M}$ are obtained by differentiation of equation (18.1):

$$d\mathbf{M} = R\,d\mathbf{I}_{n+1} = R\,(\omega_{n+1}^1 \mathbf{I}_1 + \ldots + \omega_{n+1}^n \mathbf{I}_n).$$

Thus the components of infinitesimal displacement of the point **M** are

$$\omega^i = R\omega^i_{n+1},$$ (18.2)

from which it follows that

$$\omega^i_{n+1} = \frac{1}{R}\omega^i.$$ (18.3)

Exterior differentiation of equations (18.2) gives the following exterior quadratic equations:

$$\frac{d\omega^i}{R} = \omega^k_{n+1} \wedge \omega^i_k = \frac{1}{R}\omega^k \wedge \omega^i_k.$$

Thus,

$$d\omega^i = \omega^k \wedge \omega^i_k.$$ (18.4)

Similarly,

$$d\omega^j_i = \omega^k_i \wedge \omega^j_k + \omega^{n+1}_i \wedge \omega^j_{n+1},$$

so from (18.3) and $\omega^{n+1}_i = -\omega^i_{n+1}$, we have

$$d\omega^j_i = \omega^k_i \wedge \omega^j_k - \frac{1}{R^2}\omega^i \wedge \omega^j.$$ (18.5)

Hence, the curvature of the space is

$$K = \frac{1}{R^2}.$$

With the help of (18.1) and (18.2), we can easily find that

$$d\mathbf{M} = \omega^i \, \mathbf{I}_i.$$ (18.6)

Other equations of infinitesimal displacement are

$$d\mathbf{I}_\alpha = \omega^\beta_\alpha \, \mathbf{I}_\beta.$$

In addition, again using equations (18.1) and (18.2), we find that

$$d\mathbf{I}_i = \omega^k_i \, \mathbf{I}_k + \omega^{n+1}_i \, \mathbf{I}_{n+1} = \omega^k_i \, \mathbf{I}_k - \frac{\omega^i}{R^2}\, \overrightarrow{OM},$$

or

$$d\mathbf{I}_i = \omega^k_i \, \mathbf{I}_k - K\,\omega^i \, \overrightarrow{OM}.$$ (18.7)

We will call an *analytic point* a set of coordinates

$$x^1, x^2, \ldots, x^{n+1}.$$

The *scalar square of this point* is

$$(x^1)^2 + (x^2)^2 + \ldots (x^{n+1})^2 = \mathbf{M}^2 = \frac{1}{K}.$$

Thus, we have the following *scalar products*:

$$\mathbf{M}^2 = \frac{1}{K}, \quad (\mathbf{I}_i)^2 = 1, \quad \overrightarrow{OM} \cdot \mathbf{I}_i = 0, \quad \mathbf{I}_i \cdot \mathbf{I}_j = 0. \tag{18.8}$$

The differentials $d\mathbf{M}$ and $d\mathbf{I}_i$ allow us to find the differentials of any vector considered as a point having a certain scalar square: $\mathbf{X} = X^i \mathbf{I}_i$.

108. Mapping onto an n-dimensional projective space. In order to interpret the concept of negative curvature, it is convenient to consider $n + 1$ numbers as homogeneous coordinates of a point of an elliptic space of dimension n. Here the points $(x^1, x^2, \ldots, x^{n+1})$ and $(-x^1, -x^2, \ldots, -x^{n+1})$ are considered identical.[1] Thus, an elliptic space composes only half of a spherical space.

Change the homogeneous coordinates in the space and suppose that

$$(x^1)^2 + (x^2)^2 + \ldots (x^{n+1})^2 = F\left(x^{1'}, x^{2'}, \ldots, x^{(n+1)'}\right)$$

and

$$ds^2 = F\left(dx^{1'}, dx^{2'}, \ldots, dx^{(n+1)'}\right),$$

where F is a homogeneous second-degree polynomial (a form).

As a result, we obtain the definition of the space by means of the quadratic form F satisfying the condition

$$F\left(x^{1'}, x^{2'}, \ldots, x^{(n+1)'}\right) = \frac{1}{K}. \tag{18.9}$$

We started from a definite form, but the form F could be also indefinite provided that ds^2 is positive. In this way we arrive at a hyperbolic ($K < 0$) space. We prove that this is possible only if the form F is a sum of n squares with positive signs and one square with negative sign.

A form F must be positive for all values dx related by the condition

$$\frac{\partial F}{\partial x^1} dx^1 + \cdots + \frac{\partial F}{\partial x^{n+1}} dx^{n+1} = 0. \tag{18.10}$$

This equation is obtained by differentiation of equation (18.9) and shows that the point $(dx^1, \ldots, dx^{n+1})$ lies in the polar hyperplane of the point $(x^1, \ldots, x^{n+1})$ with respect to the absolute

$$F = 0. \tag{18.11}$$

If the polar hyperplane intersects the absolute, then the differential form

$$ds^2 = F\left(dx^1, dx^2, \ldots, dx^{n+1}\right) \tag{18.12}$$

---

[1]For $n = 2$, the elliptic plane is obtained as the projection of a sphere from its center onto a tangent plane.

changes sign.

In a three-dimensional space, the last requirement excludes the ruled surfaces and admits only two cases:

1°. *The absolute is an imaginary quadric.* The form F is positive definite. The curvature K is positive. The space is elliptic.

2°. *The absolute is a real nonruled quadric,* for example, an ellipsoid. In order for the polar plane of a point to have no common points with the quadric, one must take a point (pole) inside the ellipsoid. The curvature K of the space is negative. The space is hyperbolic. Thus, a hyperbolic space is represented by interior points of the absolute. The form F decomposes into three squares with plus signs and one square with a minus sign.

109. Hyperbolic space. Let us show that for an n-dimensional hyperbolic space, we can choose the form F composed of n squares with plus signs and one square with a minus sign.

Consider a point $\mathbf{P}(0, \dots , 0, 1)$. In order for the coordinates of this point to satisfy equation (18.9), it is sufficient to take F of the form

$$F = \frac{(x^{n+1})^2}{K} + 2a_i x^i x^{n+1} + a_{ij} x^i x^j.$$

If we change the coordinate x^{n+1} by taking a new coordinate

$$x^{n+1} + K a_i x^i,$$

then the coordinates of the point $\mathbf{P}$ are not changed. Assuming that we made this coordinate transformation, we have

$$F = \frac{(x^{n+1})^2}{K} + \Phi(x^1, \dots , x^n).$$

At the point $\mathbf{P}$, linear equation (18.10) reduces to one term

$$dx^{n+1} = 0,$$

and line element (18.12) at the point $\mathbf{P}$ is positive if

$$\Phi(dx^1, \dots , dx^n) > 0.$$

Assuming that the form Φ is positive definite, we arrive at elliptic geometry if $K > 0$ and at hyperbolic geometry if $K < 0$.

Moreover, the points of the hyperbolic space are represented only by those points of the Euclidean space whose coordinates satisfy the inequality

$$F = \frac{(x^{n+1})^2}{K} + \Phi(x^1, \dots , x^n) < 0,$$

i.e., by the interior points of the absolute.

110. Representation of vectors in hyperbolic geometry. Let Q be a hyperquadric of the absolute, and let $\mathbf{M}(x^i)$ and $\mathbf{M}'(x^i + dx^i)$ be two internal neighboring points.

We can consider dx^i as coordinates of a point satisfying equation (18.10) which shows that the point (dx^i) is located in the polar hyperplane of the point $\mathbf{M}$ with respect to the absolute.

We have the same result for the velocity vector $\frac{d\mathbf{X}}{dt}$. The square of the length of the vector X^i is

$$F\left(x^1, x^2, \ldots, x^{n+1}\right).$$

This is the scalar square of the point representing the vector. Thus, *any vector is represented by a point situated in the polar hyperplane of the initial point of the vector with respect to the absolute. The scalar square of this point is equal to the scalar square of the vector.*

The scalar product of two vectors with a common initial point is equal to the scalar product of the points representing these vectors.

Two vectors are orthogonal if the points representing them are conjugate, i.e., if the directions of these two vectors are conjugate with respect to the absolute.

The case $n = 2$. There are infinitely many straight lines through a point that are parallel to a given straight line, i.e., there are infinitely many straight lines that do not have common points inside the absolute. Two perpendiculars to a straight line are parallel, and so on.

In the elliptic interpretation, for the vector

$$\mathbf{X} = X^i \mathbf{I}_i,$$

we have

$$d\mathbf{X} = dx^i\,\mathbf{I}_i + x^i\,d\mathbf{I}_i = dx^i\,\mathbf{I}_i + x^i\left(-K\,\omega^i\,\overrightarrow{OM} + \omega_i^k\,\mathbf{I}_k\right)$$

$$= -K\,\omega_i\,X^i\,\overrightarrow{OM} + \mathbf{I}_i\left(dX^i + \omega_k^i X^k\right) = -K\,\omega_i\,X^i\,\overrightarrow{OM} + \mathbf{I}_i\,DX^i.$$

It follows that

$$d\mathbf{X} = D\mathbf{X} - K\,\omega_i\,X^i\,\overrightarrow{OM}. \tag{18.13}$$

111. Geodesics in Riemannian manifold. The *acceleration* is the absolute derivative of the velocity vector with respect to time.

As usual, denote by ω^i the components of infinitesimal displacement of a moving point, by t the time, and by v^i the components of the velocity vector. We have

$$\omega^i = v^i\,dt.$$

It follows that the components g^i of the acceleration vector are

$$g^i = \frac{Dv^i}{dt} = \frac{dv^i}{dt} + v^k\,\frac{\omega_k^i}{dt}.$$

Setting

$$\omega_i^j = \gamma_i{}^j{}_k \omega^k,$$

we find that

$$g^i = \frac{dv^i}{dt} + \gamma_k{}^i{}_h v^k v^h.$$

For a geodesic, we have $g^i = 0$. It follows that the equations of geodesics are

$$\begin{cases} \dfrac{dv^i}{dt} + \gamma_k{}^i{}_h v^k v^h = 0, \\[2mm] \omega^i = v^i \, ds; \end{cases} \qquad (18.14)$$

here we set $ds = dt$.

REMARK. Consider a holonomic system with n parameters $q^1, \ldots, q^n$. The kinetic energy T is defined by the equation

$$T \, dt^2 = g_{ij} \, dq^i \, dq^j. \qquad (18.15)$$

We assume that none of the constraints depend on time.

It is easy to see that instantaneous motions of the system (without exterior forces) are represented by geodesics of the space $q^1, \ldots, q^n$.

Form (18.15) was considered by Volterra.[2] He calls v^i the characteristics of the motion.

A state of the system at a given moment is defined by the quantities $(q^i, \dot{q}^i)$, where $\dot{q}^i = \dfrac{dq^i}{dt}$. Volterra replaces $\dot{q}^i$ by linear combinations of v^i.

In particular, Volterra considers the case

$$\sum_i (v^i)^2.$$

This is our point of view.

Volterra poses the following problem: is it possible to choose v^i in such a way that $\gamma_k{}^i{}_h$ are constants (a system with independent characteristics)? If this is possible, then the first group and the second group of equations (18.14) can be integrated separately.

EXAMPLE: the Euler equations (the rotation of a rigid body). It would be interesting to study Riemannian manifolds with constant $\gamma_k{}^i{}_h$.

112. Pseudoequipollent vectors: pseudoparallelism. Is it possible to choose a rectangular frame at every point of a Riemannian manifold in such a way that

$$g^i = \frac{dv^i}{dt} \ ?$$

If this is possible, then for each geodesic, we have

$$v^i = \text{const.}$$

---

[2]Volterra, Atti Torino, 1887.

Thus, the components of the unit tangent vector are constant along a geodesic.

We call two vectors $\overrightarrow{MX}$ and $\overrightarrow{M'X'}$ *pseudoequipollent* if they have the same components with respect to the frames associated with them.

The unit vectors tangent to a geodesic are pseudoequipollent.

Suppose that two vectors tangent to two geodesics are pseudoequipollent at two points. Then they are also pseudoequipollent everywhere. We call such geodesics *pseudoparallel*.

This implies the following:

1°. *Each geodesic is pseudoparallel to itself.*

2°. *Two geodesics pseudoparallel to a third geodesic are pseudoparallel to one another.*

3°. *Through any point, there passes one and only one geodesic which is pseudoparallel to a given geodesic.*

4°. *If two geodesics intersect each other, then two other geodesics, that are pseudoparallel at their common point to the first two geodesics, intersect each other at the same angle.*

We show that the converse is also true: if it is possible to define a pseudoparallelism satisfying the following two (less strong) conditions:

1°). Through any point, there passes one and only one geodesic corresponding to a given geodesic,

2°). Two geodesics corresponding to two intersecting geodesics intersect each other at the same angle as the first pair,

then it is possible to find a frame giving $v^i = $ const.

Proof. At the point O, take an orthonormal frame. To each direction emanating from a point M, associate the components a_1, a_2, a_3 of the unit vector tangent to the corresponding geodesic at the point O. The angle of two such directions (a) and (b) is defined by the formula

$$\cos \varphi = a_1 b_1 + a_2 b_2 + a_3 b_3.$$

Thus, there exists a certain rectangular frame at the point M such that if the direction cosines of the geodesic are a_1, a_2, and a_3, then the direction cosines of the corresponding geodesic at the point O are also a_1, a_2, and a_3.

Therefore, to any geodesic emanating from the point O, there corresponds also a geodesic emanating from the point M. The components of the velocity vector of a moving point along the geodesic are also constants. It follows that

$$\gamma^i_{kh} v^k v^h = 0.$$

In fact, at an arbitrary point M_0, we have

$$\frac{d\overset{\circ}{v}{}^i}{dt} = 0.$$

Using (18.14), the last equations can be written as

$$\overset{\circ}{\gamma}^i_{kh}\overset{\circ}{v}^k\overset{\circ}{v}^h = 0.$$

It follows that two geodesics are pseudoparallel if the same geodesic emanating from the point O corresponds to them.

In Euclidean space we have an ordinary parallelism. We will see that in elliptic space, there are two types of pseudoparallelism.

Consider a rigid body moving with an angular velocity (p, q, r) and the moments $A = B = C = 1$. For the axes connected with this rigid body, we have

$$\frac{dp}{dt} = \frac{dq}{dt} = \frac{dr}{dt} = 0, \quad T = p^2 + q^2 + r^2.$$

It follows that the parallelism is "absolute".

The equations remain the same for fixed axes in the space.

THE FRENET FORMULAS FOR GEODESICS.

If $\mathbf{T}$ is the unit tangent vector at a point M, then

$$\frac{d\mathbf{M}}{ds} = \mathbf{T}, \quad \frac{d\mathbf{T}}{ds} = 0, \tag{18.16}$$

113. Geodesics in spaces of constant curvature. From (18.13), we have

$$d\mathbf{X} = D\mathbf{X} - K\, X^i \omega_i\, \overrightarrow{OM},$$

$$X^i \omega_i = \mathbf{X} \cdot d\mathbf{M}.$$

For $\mathbf{X} = \mathbf{T}$, by (18.16) and $\mathbf{T} \cdot d\mathbf{M} = ds$, we find that

$$d\mathbf{T} = -K\, ds\, \overrightarrow{OM}.$$

This implies that

$$\frac{d\mathbf{T}}{ds} = -K\, \overrightarrow{OM}.$$

Set $\overrightarrow{OM} = \mathbf{M}$. Excluding $\mathbf{T}$ by means of the first equation of (18.16), we obtain

$$\frac{d^2\mathbf{M}}{ds^2} + K\mathbf{M} = 0. \tag{18.17}$$

Integrating this equation as a linear ordinary differential equation with constant coefficients, we obtain for $K > 0$

$$\mathbf{M} = \mathbf{A}\, \cos\!\left(s\sqrt{K}\right) + \mathbf{B}\, \sin\!\left(s\sqrt{K}\right),$$

where $\mathbf{A}$ and $\mathbf{B}$ are constant points.

The point $\mathbf{M}$ describes the straight line AB.

Thus, the geodesics of an elliptic space are ordinary straight lines.

If $\mathbf{M}_0$ is an initial point of a straight line and $\mathbf{T}_0$ is the point representing an initial direction, then we easily obtain

$$\mathbf{M} = \mathbf{M}_0 \cos\left(s\sqrt{K}\right) + \frac{\mathbf{T}_0}{\sqrt{K}} \sin\left(s\sqrt{K}\right). \tag{18.18}$$

The denominator in the second term of the right-hand side is explained by the fact that according to formulas (18.8), the point $\mathbf{M}$ as all analytic points has the norm $\frac{1}{\sqrt{K}}$ while $\mathbf{T}_0$ is the unit vector.

THE TOTAL LENGTH OF A STRAIGHT LINE

1°. *Spherical space:* $K = \frac{1}{R^2}$.

$$S_T = \frac{2\pi}{\sqrt{K}} = 2\pi R. \tag{18.19}$$

Since the trigonometric functions in equation (18.18) are of period 2π, the argument $s\sqrt{K}$ can vary from 0 to 2π. Thus, the arc length s can vary from 0 to $\frac{2\pi}{\sqrt{K}}$.

2°. *Elliptic space:*

$$S_T = \frac{\pi}{\sqrt{K}}. \tag{18.20}$$

In an elliptic space, the points with coordinates of opposite signs are identified. Since if the argument is increased by π, both trigonometric functions in equation (18.18) change their signs, the period for the point $\mathbf{M}$ is π for $s\sqrt{K}$ and $\frac{\pi}{\sqrt{K}}$ for s.

We can multiply both members of equation (18.18) by $\mathbf{M}_0$. Since

$$\mathbf{M}_0 \cdot \mathbf{T}_0 = 0, \quad (\mathbf{M}_0)^2 = \frac{1}{\sqrt{K}},$$

we obtain

$$\mathbf{M}_0 \cdot \mathbf{M} = \frac{1}{K} \cos\left(s\sqrt{K}\right) = R^2 \cos\frac{s}{R}.$$

It follows that for a fixed point $\mathbf{M}_0$, the period for the point $\mathbf{M}$ is $\frac{\pi}{R}$.

3°. *Hyperbolic space:* $K = -\frac{1}{R^2} < 0$. Now when we integrate equations (18.17), we obtain hyperbolic functions instead of trigonometric ones:

$$\mathbf{M} = \mathbf{M}_0 \cosh\left(s\sqrt{-K}\right) + \frac{\mathbf{T}_0}{\sqrt{-K}} \sinh\left(s\sqrt{-K}\right)$$

$$= \mathbf{M}_0 \cosh\left(\frac{s}{R}\right) + \mathbf{T}_0 R \sinh\left(\frac{s}{R}\right). \tag{18.21}$$

If we change the normalization of the analytic point $\mathbf{M_0}$, we can write (18.21) in the form

$$\mathbf{M} = \mathbf{M_0} + \mathbf{T_0} \tanh\left(\frac{s}{R}\right). \qquad (18.22)$$

When s changes from 0 to ∞, the point $\mathbf{M}$ changes from $\mathbf{M_0}$ to $\mathbf{M_0} + \mathbf{T_0} \tanh(\frac{s}{R})$. The last point lies on the absolute since

$$(\mathbf{M_0} + \mathbf{T_0}R)^2 = (\mathbf{M_0})^2 + R^2 = 0.$$

Thus, the absolute is the locus of the points at infinity of straight lines. Each straight line has two points at infinity, $\mathbf{P}(\infty)$ and $\mathbf{Q}(-\infty)$. When the argument s changes sign, the hyperbolic tangent changes sign too, but in a hyperbolic space there is no identification of points with opposite signs of their coordinates.

114. The Cayley metric. Cayley defined the projective distance M_0 by means of the cross-ratio

$$(\mathbf{M}, \mathbf{M_0}; \mathbf{Q}, \mathbf{P}).$$

Since parameters of these four points are proportional to the numbers

$$\tanh\left(\frac{s}{R}\right), 0, -1, +1,$$

we have

$$(\mathbf{M}, \mathbf{M_0}; \mathbf{Q}, \mathbf{P}) = \frac{1 + \tanh\left(\frac{s}{R}\right)}{1 - \tanh\left(\frac{s}{R}\right)} = \frac{\sinh\left(\frac{s}{R}\right) + \cosh\left(\frac{s}{R}\right)}{-\sinh\left(\frac{s}{R}\right) + \cosh\left(\frac{s}{R}\right)} = e^{\frac{2s}{R}}.$$

It follows that

$$s = \frac{R}{2} \ln(\mathbf{M}, \mathbf{M_0}; \mathbf{Q}, \mathbf{P}).$$

Remark. In Euclidean space, take the Euclidean sphere of radius R as the absolute.

As a parameter of a point $\mathbf{M}$, we can take the Euclidean length $r = |OM|$. It is easy to derive that

$$r = \tanh\left(\frac{s}{R}\right)$$

and

$$\overrightarrow{MM_0} = -\frac{1}{R^2} \sinh\left(\frac{s}{R}\right).$$

D. THE THEORY OF GEODESIC LINES

Chapter 19

Variational Problems for Geodesics

115. The field of geodesics. We consider a family of lines close to an arc AB of a curve. Suppose that α is a parameter of the family and that $\alpha = 0$ for the curve AB.

On each line of the family, the points are defined by a parameter t. Take the unit vector tangent to an arc of a curve of the family as the coordinate vector $\mathbf{I}_1$. Choose the following symbols of differentiation:

$$d = \frac{\partial}{\partial t} \cdot dt, \quad \delta = \frac{\partial}{\partial \alpha} \cdot d\alpha,$$

$$\omega^i(d) = \omega^i, \quad \omega^i(\delta) = e^i.$$

By our choice of axis $\mathbf{I}_1$, we have

$$\omega^1 = ds, \quad \omega^2 = \ldots = \omega^n = 0.$$

From structure equations (17.1), it follows that

$$\begin{cases} d\omega^1 = \omega^i \wedge \omega_i^1, \\ d\omega^i = \omega^1 \wedge \omega_1^i + \omega^k \wedge \omega_k^i, \quad i, j, k = 2, \ldots, n, \\ d\omega_1^i = \omega_1^k \wedge \omega_k^i + R_{1i,1k}\, \omega^1 \wedge \omega^k + R_{1i,jk}\, \omega^j \wedge \omega^k, \end{cases}$$

and this implies the following bilinear decompositions

$$\begin{cases} \delta\omega^1 = de^1 + e^i \omega_i^1, \\ \delta\omega^i = de^i + e^1 \omega_1^i - e_1^i \omega^1 + e^k \omega_k^i, \\ \delta\omega_1^i = de_1^i + e_1^k \omega_k^i - e_k^i \omega_1^k - R_{1i,1k}\, e^k \omega^i. \end{cases} \tag{19.1}$$

116. Stationary state of the arc length of a geodesic in the family of lines joining two points. *We prove that the length of the arc AB in the family of lines passing through the points A and B is stationary only for a geodesic.*

Since now all lines of the family pass through the points A and B, the variation $\omega^i(\delta), i = 1, \ldots, n$, at these points vanishes:

$$e^i = 0 \text{ at the points } A \text{ and } B. \tag{19.2}$$

Since the differential of the arc length of a curve of the family is equal to ω^i, the length of this curve between the points A and B is expressed by the integral

$$\lambda = \int_A^B \omega^1.$$

It follows that under the passage from one curve of the family to another and for fixed ends A and B, the variation of this line is

$$\delta\lambda = \int_A^B \delta\omega^1.$$

By means of equations (19.1), we can write this variation as

$$\delta\lambda = \int_A^B (de^1 + e^i\omega_i^1) = \int_A^B de^1 + \int_A^B e^i\omega_i^1 = \int_A^B (e^2\omega_2^1 + \ldots + e^n\omega_n^1), \tag{19.3}$$

since the first integral $\int_A^B de^1$ reduces to the difference of the values of e^1 at the points A and B, and conditions (19.2) show that this difference is equal to zero.

In order to have $\delta\lambda = 0$ for any variations e^i, $i = 2, \ldots, n$, vanishing at the points A and B, it is necessary and sufficient that

$$\omega_1^i = 0, \quad i = 2, \ldots, n. \tag{19.4}$$

Equations (19.4) show that the arc AB in question is a geodesic. In fact, applying (15.24), we find that the absolute differential of the tangent $\mathbf{I}_1$ is

$$D\mathbf{I}_1 = \omega_1^i \mathbf{I}_i,$$

and under conditions (19.4), the tangent preserves its direction. Thus, we made our definition of a geodesic to be consistent with that of Riemann (see No. **61**).

Remark. We see that it is always possible to choose frames at points of a geodesic in such a way that

$$\omega^i = 0, \quad \omega_1^i = 0, \quad i = 2, \ldots, n.$$

117. The first variation of the arc length of a geodesic. *Compute the variation of the arc length of the geodesic AB, when AB is transferred to A′B′.*

Now we wave the requirement that all lines of the family (α) pass through the points A and B common to all of them. Instead of this requirement, we request that all lines of the family (α) are geodesics. Then condition (19.4) is extended to all curves of the family (α).

Now the variation (19.3) of the arc length of a geodesic takes the form

$$\delta\lambda = \int_A^B (de^1 + e^i\omega_i^1) = \int_A^B de^1 = (e^1)_B - (e^1)_A.$$

Here $(e^1)_A$ is the projection of the displacement AA' onto the tangent at the point A of the arc AB, and $(e^1)_B$ is the projection of the displacement BB' onto the tangent at the point B of the arc AB (Fig. 7).

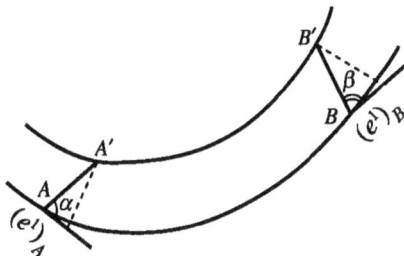

Fig. 7

Setting the angle between AA' and the tangent to AB at the point A equal to α and the angle between BB' and the tangent to AB at the point B equal to β,

$$\alpha = (\widehat{AA', AB}), \quad \beta = (\widehat{BB', AB}),$$

we obtain the desired variation of the arc length in the form

$$\delta\lambda = -AA' \cos\alpha - BB' \sin\beta.$$

COROLLARY. Consider a family of geodesics without common points. Suppose that their initial points A make up a line orthogonal to geodesics, and that all geodesics have the same length.

Then $\delta\lambda = 0$, since the length of a geodesic is not changed, and $\cos\alpha = 0$ because of orthogonality. Thus, $\cos\beta = 0$, and the trajectory BB' of the terminal point B is also orthogonal to the family of geodesics.

Thus, *if from the points of an arbitrary line AA′, we draw geodesics orthogonal to the line AA′ and plot the same segment on each geodesic, then the locus of these segments is also an orthogonal trajectory of the family of geodesics.*

As a result, we come to the notion of *parallel curves AA′ and BB′*. The notion of *parallel surfaces* (of different dimensions) can be obtained in a similar way.

118. The second variation of the arc length of a geodesic. *Compute the second variation of the arc length of a geodesic for fixed points A and B.*

Consider a continuously differentiable family of lines (α) joining the points A and B and situated in a neighborhood of the geodesic $\alpha = 0$.

Now equation (19.3) takes the form

$$\delta\lambda = \int_A^B de^1 + \int_A^B e^i \omega_i^1. \tag{19.5}$$

We cannot assume that $\omega_1^i = 0$, since except for the line $\alpha = 0$, the lines of the family (α) are not geodesics. But for fixed ends A and B, we have

$$\omega^i(\delta) = e^i = 0 \quad \text{at the points} \quad A \quad \text{and} \quad B. \tag{19.6}$$

Therefore, the first integral on the right-hand side of (19.5) vanishes, and

$$\delta\lambda = \int_A^B e^i \omega_i^1.$$

For fixed ends A and B, we have

$$\delta^2\lambda = \int_A^B (\delta e^i \, \omega_i^1 + e^i \, \delta\omega_i^1).$$

But for the geodesic $\alpha = 0$, we have $\omega_i^1 = 0$, and

$$\delta^2\lambda = \int_A^B e^i \, \delta\omega_i^1. \tag{19.7}$$

Additional conditions. Suppose that at every point of the geodesic AB, a frame is obtained from an initial one at the point A by a *parallel transport*. This implies that along the geodesic AB, the forms ω_i^j vanish,

$$\omega_i^j = 0, \quad i, j = 1, \dots, n. \tag{19.8}$$

Thus, using the third equation of (19.1) and (19.8), we obtain

$$\delta\omega_1^i = de_1^i - R_{1i,1k} \, e^k \, \omega^1,$$

and as $\omega_i^1 = -\omega_1^i$, $\omega^1 = ds$, equation (19.7) takes the form

$$\delta^2\lambda = \int_A^B (-e_i \, de_1^i + R_{1i,1k} \, e^i \, e^k ds). \tag{19.9}$$

But by means of

$$\omega^1 = ds, \quad \omega^2 = \ldots = \omega^n = 0, \quad \delta\omega^i = 0,$$

and equations (19.8), the second equation of (19.1) gives

$$de^i = e_1^i ds.$$

Now we can integrate by parts the first terms on the right-hand side of equation (19.9):

$$-\int_A^B e^i \, de_1^i = -e^i \, e_1^i \Big|_A^B + \int_A^B (e_1^i)^2 \, ds = \int_A^B (e_1^i)^2 \, ds,$$

since $e_i = 0$ at the points A and B.

Hence, by equation (19.9), the second variation takes the form

$$\delta^2 \lambda = \int_A^B \left[\sum_i (e^i)^2 + R_{1i,1k} \, e^i \, e^k \right] ds. \tag{19.10}$$

Consider now the Riemannian curvature in the planar direction defined by the geodesic AB and the vector $\overrightarrow{MM'}$, where M and M' correspond to the same value of t on two lines of the family. The matrix of components defining these directions is

$$\begin{pmatrix} 1 & 0 & 0 & \ldots & 0 \\ e^1 & e^2 & e^3 & \ldots & e^n \end{pmatrix},$$

and the Riemannian curvature (16.13) is

$$K = -\frac{R_{1i,1k} \, e^i \, e^k}{\sum_i (e^i)^2}, \quad i \neq 1. \tag{19.11}$$

Equations (19.10) and (19.11) imply that

$$\delta^2 \lambda = \int_A^B \left[\sum_i (e_1^i)^2 - K \sum_i (e^i)^2 \right] ds. \tag{19.12}$$

If the curvature is negative, $K < 0$, then $\delta^2 \lambda > 0$, and the arc of the geodesic has its minimum (this conclusion is not rigorous). If $K > 0$ and the arc AB is not too long, then again $\delta^2 \lambda > 0$.

119. The minimum for the arc length of a geodesic (Darboux's proof). Consider a hypersurface through the point A which is normal to the arc AB and a family of parallel hypersurfaces passing through the points of the arc AB.

If the arc AB is not too long, then through every point near the arc AB there passes one and only one of these hypersurfaces. In order to define a point P lying on a hypersurface of the family passing through a point M of the line AB, as coordinates we take the parameter

$$u = AM$$

and the parameters $v, w, \ldots$ of the hypersurface (M) with line element $d\sigma^2$. Then for the line passing through the point P and joining the points A and B, we obtain the line element

$$ds^2 = du^2 + d\sigma^2.$$

Thus, we have

$$\int_0^{u_1} \sqrt{du^2 + d\sigma^2} > \int_0^{u_1} du,$$

i.e.,

$$AP > AM,$$

and the geodesic $\alpha = 0$ is the shortest line.

120. Family of geodesics of equal length intersecting the same geodesic at a constant angle. *Suppose that we have a family of geodesics of the same length a emanating from the points of the geodesic AB and making with AB the same angle φ. Let us find the difference between the length of AB and the length of the line formed by the ends of geodesics of the family, if the segment a is infinitesimally small.*

For $\alpha = 0$, taking $\delta = \dfrac{\partial}{\partial \alpha}$ and $\delta a = 1$, we have

$$e^1 = \cos\varphi, \tag{19.13}$$

and for an appropriate choice of frame, we also have

$$e^2 = \sin\varphi, \quad e^3 = \ldots = e^n = 0. \tag{19.14}$$

(Obviously, the choice of frame here is different from that in No. **118**.) For displacements along geodesics of our family (for increasing α), the variation of the arc length of AB is determined by the formula (19.5),

$$\delta\lambda = \int_A^B de^1 + \int_A^B e^i \omega_i^1, \tag{19.15}$$

or

$$\delta\lambda = (e^1)_B - (e^1)_A + \int_A^B e^i \omega_i^1. \tag{19.16}$$

The second variation is

$$\delta^2\lambda = (\delta e^1)_B - (\delta e^1)_A + \int_A^B e^i\,\delta\omega_i^1.$$

Applying formulas (19.13) and (19.14) and taking into account that for an orthonormal frame, $\omega_{\bar{1}}^1 = 0$, we find that

$$\delta^2\lambda = (\delta e^1)_B - (\delta e^1)_A + \int_A^B e^2\,\delta\omega_2^1.$$

For the geodesic $\alpha = 0$, by means of (19.4), the form ω_1^n vanishes. Thus, since $\delta\omega_2^1 = -\delta\omega_1^2$, from the third equation of (19.1), we have

$$\delta\omega_2^1 = -de_1^2 - e_1^k\omega_k^2 + R_{12,1i}\,e^i\,\omega^1.$$

Thus,

$$\delta^2\lambda = (\delta e^1)_B - (\delta e^1)_A - \sin\varphi\int_A^B (de_1^2 + e_1^k\omega_k^2 - R_{12,1i}\,e^i\,\omega^1). \qquad (19.17)$$

But for $i = 2$, the second equation of (19.1) gives

$$0 = -e_1^2\,ds,$$

since on the line $\alpha = 0$, we have $\omega^2 = 0$, $e^2 = \sin\varphi = \text{const.}$, $\omega_1^i = 0$, $\omega_k^i = 0$, and for $i > 2$, it gives

$$0 = -e_1^i\,ds + a_2^i\sin\varphi\,ds,$$

since for $i > 2$, we have $\delta\omega^i = 0$, $e^i = 0$, $\omega_2^i = a_2^i\,ds$. It follows that

$$e_1^2 = 0, \quad e_1^i = a_2^i\sin\varphi, \quad i > 2.$$

Thus, the second variation (19.17) takes the form

$$\delta^2\lambda = (\delta e^1)_B - (\delta e^1)_A + \sin\varphi\int_A^B\left(\sum_i(a_2^i)^2\sin\varphi\,ds + R_{12,12}\sin\varphi\,ds\right)$$
$$= (\delta e^1)_B - (\delta e^1)_A + \sin^2\varphi\int_A^B\left(\sum_i(a_2^i)^2 - K\right)ds,$$

where K is the Riemannian curvature in the planar direction (AB, AA').
    For our construction of a frame,

$$De^i = \delta e^i + e^k e_k^i = 0,$$

and since the line AA' is a geodesic, then $e_k^1 = 0$ and

$$\delta e^1 = 0.$$

Hence,

$$\delta^2 \lambda = \sin^2 \varphi \int_A^B \left(\sum_i (a_2^i)^2 - K \right) ds. \tag{19.18}$$

But

$$d\mathbf{I}_2 = \omega_2^i \, \mathbf{I}_i = a_2^i \, \mathbf{I}_i \, ds,$$

$$|d\mathbf{I}_2| = ds \sqrt{\sum_i (a_2^i)^2}.$$

Thus,

$$\sum_i (a_2^i)^2 = \frac{1}{h^2},$$

where h is a reduced pitch of a helical motion moving AA' to a neighboring location.

Therefore,

$$\delta^2 \lambda = \sin^2 \varphi \int_A^B \left(\frac{1}{h^2} - K \right) ds \approx \left(\frac{1}{h^2} - K \right) l, \tag{19.19}$$

where

$$l = \int_A^B ds = AB.$$

Now decomposing the length

$$A'B' = \lambda$$

into a series with respect to powers of the infinitesimal length a plotted on geodesics of our family, we obtain

$$\lambda = l + a\,\delta\lambda + \frac{1}{2}a^2\,\delta^2\lambda + \dots . \tag{19.20}$$

Since for the geodesic $\alpha = 0$ the first variation $\delta\lambda = 0$, we have

$$\lambda = l + \frac{1}{2}a^2 l \sin^2 \varphi \left(\frac{1}{h^2} - K \right) \tag{19.21}$$

and

$$\lambda^2 = l^2 + a^2 l^2 \sin^2 \varphi \left(\frac{1}{h^2} - K \right).$$

If we denote by S the area of the "parallelogram" $ABB'A'$ (Fig. 8), from the last equation we find that

$$\frac{l^2 - \lambda^2}{S^2} = K - \frac{1}{h^2}. \tag{19.22}$$

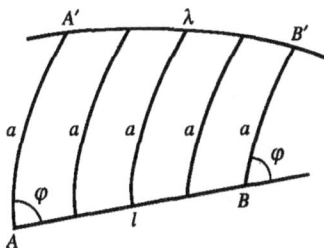

Fig. 8

1°. Suppose that $\dfrac{1}{h} = 0$. Then AA' and BB' are "parallel" in the sense of Levi–Civita. Levi-Civita called the "*parallelogramoid*" the figure obtained if one draws two "parallel" geodesics AA' and BB' of the same length through the ends A and B of the geodesic AB and join the points A' and B' by the arc $A'B'$ of another geodesic.

Our "parallelogram" is a similar figure for $h^{-1} = 0$. In this case,

$$K = \frac{l^2 - \lambda^2}{S^2}. \tag{19.23}$$

In Euclidean space, we have $l = \lambda$ while in elliptic space $\lambda < l$, and in hyperbolic space $\lambda > l$.

EXAMPLE. *Two-dimensional spherical space* $\left(n = 2,\ \varphi = \dfrac{\pi}{2}\right)$. We see that the arc of a small circle $A'B' < AB$. The quadrangle $ABB'A'$ is not a parallelogramoid.

We again obtain

$$K = \frac{1}{R^2} \quad (R \text{ is the radius of the sphere}).$$

2°. Suppose that $\dfrac{1}{h} \neq 0$. If $K < 0$, then the difference

$$\lambda - l = \frac{\dfrac{1}{h^2} - K}{\lambda + l} S^2$$

is greater than the corresponding difference in Euclidean space. It can decrease in elliptic space, but it is possible to choose h in such a way that

$\lambda = l$ and even $\lambda > l$ (Fig. 9).

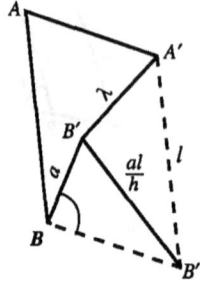

Fig. 9

EXAMPLE. $n = 3$, $K = 0$, $\varphi = \dfrac{\pi}{2}$. We again find that

$$\lambda^2 = l^2 + \frac{a^2 l^2}{h^2}.$$

Chapter 20

Distribution of Geodesics near a Given Geodesic

121. Distance between neighboring geodesics and curvature of a manifold. We return to the frame chosen in No. **118**. There we transported by parallelism a frame along a geodesic. Along a geodesic, we have

$$\omega_i^j = 0. \tag{20.1}$$

Now applying formulas (19.1), we obtain

$$\begin{cases} \delta\omega^1 = de^1, \\ 0 = de^i - e_1^i \, ds, \\ 0 = de_1^i - R_{1i,1k} \, e^k \, ds. \end{cases} \tag{20.2}$$

The above equations give us the rule which we should follow when we go off the geodesic AB in order to find an infinitesimally close geodesic.

Differentiating the second equation of (20.2) with respect to s and applying the third equation of (20.2) to eliminate $\frac{de_1^i}{ds}$, we find that

$$\frac{d^2 e^i}{ds^2} - R_1{}^i{}_{1k} e^k = 0, \quad i = 2, \ldots, n, \tag{20.3}$$

where e^i remains arbitrary.

Setting

$$R_1{}^i{}_{1k} = -A_k^i, \tag{20.4}$$

we write

$$\frac{d^2 e^i}{ds^2} + A_k^i \, e^k = 0. \tag{20.5}$$

179

The curvature in the planar direction defined by the geodesic and the arc (e^i) is

$$K = \frac{A_{ij}\, e^i\, e^j}{\sum_i (e^i)^2}.$$ (20.6)

Take a parallelogramoid $ABB'A'$ (Fig.10). Set

$$AB = l, \quad AA' = a.$$

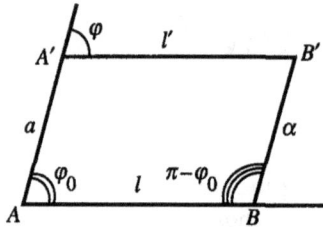

Fig. 10

We look for the upper base (a geodesic joining the points A' and B')

$$A'B' = l'.$$

Denote the angle at the point A by φ_0. Thus $\pi - \varphi_0$ is the angle at the point B.

Let λ be the length of the arc $A'B'$, the locus of points equidistant from the geodesic AB. Then from formula (19.23), we have

$$\frac{l^2 - \lambda^2}{S^2} = K.$$ (20.7)

The difference $\lambda - l'$ is negligible in our computations which only consider the terms up to $a^2 l$ (see (19.21)), since if α is the average distance between l and l', then $\lambda - l'$ is of order $\alpha^2 l'$ or $\alpha^2 l$, and α is of order $a l$. Thus, the difference $\lambda - l'$ is of order $a^2 l^3$.

The computation of formula (20.7) is straightforward.

If we set $\delta = \dfrac{\partial}{\partial a}$, then from equation (19.15) we obtain

$$\begin{cases} \delta\lambda = (e^1)_B - (e^1)_A, \\[2mm] \delta^2\lambda = (\delta e^1)_B - (\delta e^1)_A. \end{cases}$$ (20.8)

But we have

$$\delta e^1 = e^i e^1_i,$$

and from (19.13) and (19.14), at the points A and B we have

$$e^1 = \cos \varphi_0, \quad e^2 = \sin \varphi_0, \quad e^3 = \ldots = e^n = 0. \tag{20.9}$$

Therefore, substituting these values into the second equation of (20.8), we obtain

$$\delta^2 \lambda = (e_2 e_2')_B - (e_2 e_2')_A,$$

where by means of the second equation of (20.2),

$$e_i' = \frac{de^i}{ds} = e_1^i$$

and

$$\delta^2 \lambda = \sin \varphi_0 \left[(e_2')_B - (e_2')_A \right] = l \sin \varphi_0 \, (e_2'')_C,$$

where C is an appropriately chosen point between A and B, and $e_2'' = \dfrac{d^2 e}{ds^2}$.

For $\dfrac{1}{h} = 0$, by means of equation (19.19), we have

$$\delta^2 \lambda = -K l \sin^2 \varphi_0,$$

i.e., by means of (19.20),

$$l' = l - \frac{1}{2} K l \sin^2 \varphi_0,$$

and we return to the formula

$$\frac{l^2 - \lambda^2}{S^2} = K.$$

122. The sum of the angles of a parallelogramoid. Denote the angle between the directions AA' and $A'B'$ by φ. From (20.9), at the point A, we have

$$- \sin \varphi_0 \, \delta \varphi = \delta e^1 = \sin \varphi_0 \, (e_2')_A.$$

Cancelling $\sin \varphi_0$, we obtain the equation

$$\delta \varphi = -(e_2')_A.$$

Integrating this equation under the condition $\delta a = 1$ (No. **120**), we find that

$$\varphi - \varphi_0 = -a(e_2')_A. \tag{20.10}$$

Since $AB = l$, at the point B, we have

$$e_2 = (e_2)_A + l \, (e_2')_A + \frac{1}{2} l^2 \, (e_2')_A.$$

From this equation, using (19.14) and (20.9), we find that

$$\sin \varphi = \sin \varphi_0 + l \, (e_2')_A + \frac{1}{2} l^2 \, (e_2')_A.$$

Since

$$\sin\varphi - \sin\varphi_0 = (\varphi - \varphi_0)\cos\varphi_1, \quad \varphi_1 = \varphi_0 + \theta(\varphi - \varphi_0), \quad 0 \le \theta \le 1,$$

we find that

$$-a\cos\varphi_1 (e_2')_A = l(e_2')_A + \frac{1}{2}l^2 (e_2')_A.$$

Assuming that $\dfrac{a}{l}$ tends to 0, we find from the last equation that

$$(e_2')_A = -\frac{1}{2}l(e_2')_A,$$

and hence by (20.3) and $R_1{}^2{}_{12} = -K$, we have[1]

$$(e_2')_A = \frac{1}{2}Kl\sin\varphi_0. \tag{20.11}$$

Comparing equations (20.10) and (20.11), we find that

$$\varphi - \varphi_0 = -\frac{1}{2}Kal\sin\varphi_0,$$

i.e.,

$$\varphi = \varphi_0 - \frac{1}{2}KS,$$

where $S = al\sin\varphi_0$ is the area of the parallelogramoid.

The sum of the interior angles A and A' is

$$\pi + \frac{1}{2}KS,$$

and the sum of the all angles of the parallelogramoid is

$$\pi + KS.$$

These are results of Saccheri.

123. Stability of a motion of a material system without external forces. *Suppose that a geodesic $A'B'$ is infinitesimally close at the point A' to a given geodesic AB. Are all the points of $A'B'$ also infinitesimally close to the points of AB?*

In this case, it is always possible to arrange that A_{ij} (see (20.4)) have the values

$$A_{ij} = 0, \quad i \ne j, \quad A_{ii} = K_i.$$

The basic equations

$$\frac{d^2 e^i}{ds^2} + K_i e^i = 0 \quad (i \text{ not summed}) \tag{20.12}$$

---

[1] In the original, the sign " $-$ " was erroneously written on the right-hand side of the formula (20.11). (*Translator's note.*)

are now differential equations with separable variables. If one of the components $K_i < 0$, then we have an instability. If all $K_i > 0$, then the motion is stable (Fig. 11).

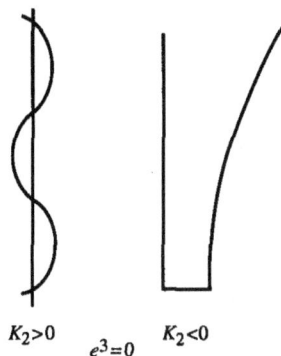

$$K_2 > 0 \qquad \qquad K_2 < 0$$
$$e^3 = 0$$

Fig. 11

Suppose that the dimension n of the manifold is three, $n = 3$. Develop a geodesic on the Euclidean space. The image is a plane curve in the preferred direction e^2 (or e^3). In other directions, the image is a space curve.

124. Investigation of the maximum and minimum for the length of a geodesic in the case $A_{ij} = $ const. Suppose that for fixed points A and B, we have

$$\delta^2 \lambda = \int_A^B \sum_i [(e_i')^2 - K_i\,(e_i)^2]\,ds.$$

Let us assume that the manifold is two-dimensional ($n = 2$) and of constant curvature. Then

$$\delta^2 \lambda = \int_A^B [(e')^2 - K\,e^2]\,ds,$$

where e satisfies the equation

$$U'' + K U = 0; \quad U'' = \frac{d^2 U}{ds^2}.$$

If the equation $U'' + K U = 0$ admits a nonvanishing solution between A and B, i.e., if there is a neighboring geodesic having no common points with AB, then

$\delta^2 \lambda > 0$, and the minimum exists (provided that some additional conditions are satisfied). Suppose that $K > 0$. We can set

$$e = \eta U, \quad \eta = 0 \text{ at the points } A \text{ and } B.$$

Then

$$\delta^2 \lambda = \int_A^B [U^2(\eta')^2 + 2U U' \eta \eta' + (U')^2 \eta^2 - K U^2 \eta^2] \, ds.$$

We have

$$\int_A^B 2U U' \eta \eta' \, ds = \int_A^B U U' \, d(\eta)^2 \, ds = [U U' \eta)^2]_A^B - \int_A^B \eta^2 \, d(U U')$$
$$= \int_A^B \eta^2 [U U'' + (U')^2] \, ds$$

and

$$\delta^2 \lambda = \int_A^B [U^2(\eta')^2 - U U' \eta^2 - K U^2 \eta^2] \, ds = \int_A^B U^2 (\eta')^2 \, ds > 0. \qquad (20.13)$$

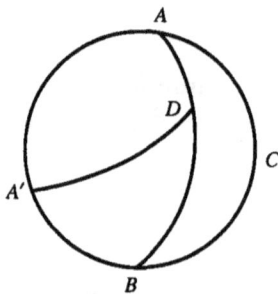

Fig. 12

This condition is satisfied if $K \leq 0$. If $K > 0$, then

$$U = a \sin(s - s_0) \sqrt{K}.$$

If the displacement of the length is less than $\dfrac{\pi}{\sqrt{K}}$, the condition $U \neq 0$ holds, and we have a minimum.

EXAMPLE OF A SPHERE. If $ACBA' > \pi$ (Fig. 12), there is no minimum for the length of a geodesic. In fact, consider a neighboring arc $ADB = ACB = \pi$, where D is an arbitrary point infinitesimally close to B. Take the arc of the great circle DA' through D. We have

$$AD + DA' < AD + DB + BA' < ACBA'.$$

It is always possible to eliminate the singular point D.

Return back to the general problem of arbitrary dimension n. A minimum is guaranteed if all $K_i \leq 0$. Otherwise, we take $\tilde{K}_i = \min K_i$. A minimum is guaranteed for a geodesic of length less than $\frac{\pi}{\sqrt{\tilde{K}_i}}$. [2]

125. Symmetric vectors. Consider a point O of the Riemannian manifold. For a given point M, consider a geodesic OM and the arc $OM' = OM$ on the prolongation of this geodesic. The point M' is called *symmetric* to the point M with respect to the point O.

Consider a vector $\mathbf{V}$ emanating from the point M. This vector is the velocity of a moving point describing a certain line (C) for which a certain line (C') is symmetric with respect to the point O. The velocity $\mathbf{V}'$ $(V' = V)$ of a moving point describing the line (C') is the vector which is symmetric to the vector $\mathbf{V}$.

We transport by parallelism the vector $\mathbf{V}$ to a vector $\mathbf{V}_1$ with its initial point at M'.

Theorem 1. *The vector $\mathbf{V}_1$ is opposite to the vector $\mathbf{V}'$ to within second-order infinitesimals.*

Proof. We have a family of geodesics MOM' depending on the parameter α. Transport by parallelism the frames from O to every point M in such a way that

$$\omega_i^j = 0.$$

Then we have the Levi–Civita formulas (20.5),

$$\frac{d^2 e^i}{ds^2} + A_k^i e^k = 0.$$

For $s = 0$, all e^i vanish. Hence, all $e^{i'}$ also vanish for $s = 0$.

Set

$$e^{i'} = a^i.$$

Up to fourth-order infinitesimals, we have

$$e^i = a^i s - \frac{1}{6} s^3 A_k^i a^k.$$

Thus, at the point M,

$$e^i(+l) = a^i l - \frac{1}{6} l^3 A_k^i a^k,$$

---

[2]Goursat, E., *Cours d'analyse; Tome 3: Intégrales infiniment voisines. Équations aux dérivées partielles du second ordre. Équations intégrales. Calcul des variations*, 3d edition, 1921. (JFM **48**, p. 491.)

and at the point M',

$$e^i(-l) = -a^i l + \frac{1}{6} l^3 A^i_k a^k.$$

So, at the points M and M', e^i have the same absolute values but opposite signs.

Set also

$$a^i l = b^i.$$

Then

$$e^i(l) = b^i - \frac{1}{6} l^2 A^i_k b^k,$$

$$e^i(-l) = -b^i + \frac{1}{6} l^2 A^i_k b^k.$$

On the other hand,

$$e^1(l) = l' - l,$$

$$-e^1(-l) = l' - l,$$

where l' is the arc length of a geodesic close to OM. Therefore, by putting

$$e^1(l) = b^i,$$

we find that

$$e^1(-l) = -b^i$$

as desired.

For finite vectors $\mathbf{V}$ and $\mathbf{V}'$, all b^i are finite. This implies the theorem since the components of $\mathbf{V}'$ are the same as those of $\mathbf{V}$.   □

126. Parallel transport by symmetry. Thus, in order to transport by parallelism a vector $\mathbf{V}$ from the point M to the point M', we construct at the point M' the vector $\mathbf{V}'$, which is symmetric to the vector $\mathbf{V}$, and take the vector $\mathbf{V}_1$ opposite to $\mathbf{V}'$. This construction is exact to within third-order infinitesimals.

The construction will be completely exact if the quantities e^i are odd functions, i.e., if the quantities A^i_k are even functions. This is the case, for example, if all A^i_k are constants (the case considered in No. **123** and No. **124**). The curvature in the direction of a plane element is preserved if one transports by parallelism this element tangentially to it. Then the symmetry preserves ds^2 (the symmetry is an *isometric transformation*).

This reduces to the fact that two moving points, that are constantly symmetric with respect to the point O, have the same velocity at all times. The latter is obvious from our previous discussion.

Conversely, suppose that the symmetry with respect to a point O is isometry. Then parallel transport preserves the curvature.

In fact, consider a bivector $(\mathbf{X}, \mathbf{Y})$ at a point M and transport by parallelism this bivector to a bivector $M'(\mathbf{X}', \mathbf{Y}')$. Let O be the midpoint of the geodesic MM', and the bivector $(\mathbf{X}_1, \mathbf{Y}_1)$ is symmetric to the bivector $(\mathbf{X}, \mathbf{Y})$. Since the symmetry is isometric, it preserves the curvature. Thus, the curvature of the area element $M'(\mathbf{X}_1, \mathbf{Y}_1)$ is equal to that of $M(\mathbf{X}, \mathbf{Y})$. But $(\mathbf{X}_1, \mathbf{Y}_1)$ coincides with $(\mathbf{X}', \mathbf{Y}')$ up to third-order infinitesimals.

Hence, the Riemannian curvature K is preserved up to third-order infinitesimals.

REMARK. It is not necessary that a parallel transport is tangent to a plane element.

PARTICULAR CASE: a space of constant curvature.

Corollary 2. *Take two points O and O' on the same geodesic and make consecutively two symmetries. Under such a transformation, the geodesic slides along itself. This transformation becomes isometry, if the symmetry possesses this property.*

127. Determination of three-dimensional manifolds, in which the parallel transport preserves the curvature. We put aside the case when the quadric-indicatrix Q of curvature is a sphere.

First, suppose that K_{11}, K_{22}, and K_{33} are mutually distinct. We prove that this case cannot take place. In fact, take at each point the trihedrons of principal directions. A principal direction remains principal under a parallel transport. Thus, the vector $\mathbf{I}_i$ is transported to the vector $\mathbf{I}'_i$, and as a result, $D\mathbf{I}_i = 0$. It follows that

$$\omega_i^j = 0.$$

Hence, the curvature vanishes, and this contradicts the hypothesis.

Only the case when Q is a surface of revolution remains to be considered. For example, we take

$$K_{11} = K_{22} = A, \quad K_{33} = C, \quad A \neq C.$$

We superpose the vector $\mathbf{I}_1$ with the axis of rotation of the indicatrix. The parallel transport moves the vector $\mathbf{I}_3$ to the vector $\mathbf{I}'_3$, but does not necessarily preserve $\mathbf{I}_1$ and $\mathbf{I}_2$.

We have only

$$D\mathbf{I}_3 = 0.$$

Therefore,

$$\omega_1^3 = 0, \quad \omega_2^3 = 0,$$

and also

$$A = \text{const.}, \quad C = \text{const.}$$

Conversely, suppose that the last four equations hold. Then the parallel transport preserves the curvature

$$A(\alpha^2 + \beta^2) + C\gamma^2,$$

since A, C, γ, and hence $\alpha^2 + \beta^2$, are preserved. So, we have the structure equations in the form

$$\begin{cases} d\omega^1 = \omega^2 \wedge \omega^1_2, \\[2mm] d\omega^2 = \omega^1 \wedge \omega^2_1, \\[2mm] d\omega^3 = 0, \\[2mm] d\omega^2_1 = -C\omega^1 \wedge \omega^2. \end{cases} \qquad (20.14)$$

Moreover, we have

$$d\omega^3_1 = -A\omega^1 \wedge \omega^3.$$

Thus,

$$A = B = 0.$$

Do manifolds exist satisfying the system of equations (20.14)?
First of all, the third equation (20.14) implies

$$\omega^3 = dz.$$

Three other equations of (20.14) are the structure equations of a two-dimensional space of constant curvature $C \neq 0$.
Thus, all solutions are

$$ds^2 = dz^2 + d\sigma^2,$$

where $d\sigma^2 \neq 0$ is a line element of a two-dimensional space of constant curvature.

We can find all isometric transformations of this space. Such a transformation must preserve the basis vector $\mathbf{I}_3$, and hence also dz:

$$dz = dz'.$$

This implies that

$$z' = z + c,$$

where $c = \text{const}.$

In addition, there are two more families of isometric transformations $d\sigma^2$ (corresponding to the proper motions and the motions with symmetry).

Altogether, there are four families of transformations.

REMARK. There are important results for the manifolds of dimension higher than three. See these results in Chapter 11 "Symmetries and parallel transport. Symmetric spaces" of the book É. Cartan, *Les leçons sur la géométrie des espaces de Riemann*, Gauthier-Villars, Facs. 2, Paris, 1963. (MR **8**, 602; Zbl. **60**, 381.); English translation: É. Cartan, *Geometry of Riemannian spaces*, Math. Sci. Press, Brookline, MA, 1983; translated by James Glazebrook. (MR 85m:53001; Zbl. 603.53001.)

Chapter 21

Geodesic Surfaces

128. Geodesic surface at a point. Severi's method of parallel transport of a vector. If through a given point A in Riemannian manifold, we take the various geodesics (of the manifold) tangent at this point to the same plane element, we obtain a surface that is called *geodesic* at the point A. To see that a given surface is geodesic at one of its points, it suffices to consider the geodesics tangent to the surface at that point: they must lie entirely on the surface.

A geodesic surface at A has at this point vanishing principal curvatures, since the normal curvature is zero when passage is made from the point A to a point A' which is infinitesimally close to A on the surface. If we then transport by parallelism from A to A' a vector tangent to the surface at A, then it remains tangent to the surface at A' (up to infinitesimals of an order greater than the first).

It is possible to derive from this a geometric construction due to Severi[1] concerning parallel transport of a vector.

Theorem 1 (Severi). *To transport by equipollence a vector* $\mathbf{X}$ *emanating from a point A to a point A', which is infinitesimally close to A, one must construct the joining geodesic AA' as well as the geodesic surface at A tangent to this geodesic and to the vector* $\mathbf{X}$ *at A. The required vector* $\mathbf{X}'$ *is tangent to this surface at A' and makes with the geodesic AA', prolonged on the other side of A', the same angle that the given vector* $\mathbf{X}$ *makes with the geodesic AA'.*

The last part of this theorem results from the fact that the direction of the geodesic AA' at the point A' is *parallel* to its direction at the point A and that parallel transport preserves the angle between two directions.

It is important to note that if we were to transport by parallelism the vector $\mathbf{X}$ along a *finite* arc of the geodesic AA' (sufficiently prolonged), the vector obtained would no longer be, in general, tangent to the considered geodesic surface, since there is no reason, *a priori*, to think that the surface, geodesic at A, is geodesic at the other points of the geodesic AA'.

---

[1] F. Severi, *Sulla curvature delle superficie a varietá*, Rend. Circ. matem. di Palermo, **42** (1917), pp. 227–229. (JFM **46**, p. 1068.)

If two surfaces, intersecting along a line (C), are geodesic at all points of (C), then they intersect at a constant angle, as the line (C) could be regarded as a curvature line on each of the surfaces.

In fact, if we develop the line (C) in Euclidean space (No. **67**), then at each point, this Euclidean space is tangent to both surfaces. The manifold of plane tangent elements of each of the surfaces forms a "curvature band" (in the sense of Blaschke). This means that after development, in a neighborhood of the developed line (C), the images of the tangent planes of both geodesic surfaces behave similarly to the tangent planes of two surfaces in Euclidean space, and the latter have a common curvature line as their intersection.

In particular, the Joachimstahl theorem on the intersection of two surfaces at a constant angle is preserved.

129. Totally geodesic surfaces. A surface which is geodesic at each of its points is called *totally geodesic*.[2] It enjoys the characteristic property that every geodesic which is tangent to this surface is completely contained within this surface.

Another characteristic property is that every geodesic that has two of its points (sufficiently close) on the surface is also completely contained within it. In fact, let A and A' be these two points. From the point A, there emanates an infinity of geodesics along the surface. They cover the whole surface within a sufficiently small neighborhood. Therefore, one of them passes through the point A', and since only one geodesic passes through two sufficiently close points in the manifold, it is precisely the geodesic in question. The converse can be proved in a similar way.

Thus, totally geodesic surfaces possess the characteristic properties of the *plane* in Euclidean space. We would reserve the name of *planes* for them. However, we will see that *the existence of such "planes" in a Riemannian manifold is an exceptional event.*

A totally geodesic surface enjoys the following three properties which are equivalent to one another:

First, *it has zero principal curvatures at each of its points.*

Second, *the unit vector normal to the surface remains normal under parallel transport along the surface.*

Third, *every vector tangent to the surface remains tangent under parallel transport along the surface.*

The second property is an immediate consequence of the first which in turn can be derived from the second. As for the second and the third, they are obviously equivalent to each other.

Let us show that the third property is *characteristic* of totally geodesic surfaces. This results in each of the first two properties being also characteristic of these surfaces.

Let us assume that every vector tangent to a given surface (S) remains tangent under parallel transport along (S), when its initial point describes an

---

[2]J. Hadamard, *Sur les éléments linéaires à plusieurs dimensions*, Darboux Bull. (2) **25** (1901), pp. 37–40. (JFM **32**, p. 614.)

arbitrary curve on (S). This implies that the acceleration of a point describing an arbitrary curve on the surface (S) is always tangent to (S). If we express the coordinates of a point on the surface in terms of functions of two parameters α and β, it will suffice to have two second-order differential equations in α and β in order to show that the acceleration of the moving point is zero. *Therefore, there exists on the surface a geodesic (of the manifold) passing through an arbitrary point of this surface and having at this point an arbitrary direction tangent to the surface.* Consequently, the surface is totally geodesic. The proof by computation is not difficult.

130. Development of lines of a totally geodesic surface on a plane. Another important property of totally geodesic surfaces is the following: *If we develop in Euclidean space any one of the curves traced on a totally geodesic surface, a plane curve is obtained.*

In fact, since the absolute differential of the unit tangent vector is tangent to the surface, the principal normal is tangent to the surface. The unit vector of the binormal is then normal to the surface, and its absolute differential is zero. Thus, the torsion of the curve is zero. So, in the development, the curve with zero torsion becomes a plane curve.

The converse is also true. Suppose that for every curve on a surface, the torsion is zero, or what amounts to the same thing, that the velocity, the first and second order accelerations of a point moving along the surface in any manner are constantly coplanar. Suppose that the surface is defined by the equation $u^3 = 0$. It is always possible to imagine another point moving along the surface in such away that at any given instance, the quantities $\frac{du^i}{dt}$ and $\frac{d^2u^i}{dt^2}$ have the same numerical values as for the first moving point, whereas the quantities $\frac{d^3u^1}{dt^3}$ and $\frac{d^3u^i}{dt^3}$ vary from the first moving point to the second by arbitrary quantities α^1 and α^2, respectively. This implies that the arbitrary vector (α^1, α^2) tangent to the surface is coplanar with the velocity and acceleration of the first point. Consequently, this acceleration lies in the tangent plane to the surface, and all the lines on the surface are asymptotic. Thus, the surface is totally geodesic.

131. The Ricci theorem on orthogonal trajectories of totally geodesic surfaces. From this, we are going to deduce a remarkable theorem due to G. Ricci.[3]

Theorem 2. *If in a Riemannian manifold there exists a one-parameter family of planes, then their orthogonal trajectories establish an isometric point correspondence between the different planes of the family.*

Proof. In fact, let (P) be a plane in the family, and let (P') be an infinitesimally close plane. Let (C) be an arbitrary curve traced in the plane (P), and let (C') be the locus of the common points of (P') and the orthogonal trajectories taken through the different points of (C). Let us construct the representation in the Euclidean space of conjugacy (No. **67**) along the curve (C). In this

---

[3]G. Ricci, *Sulle superficie geodetiche in nun varietà qualunque e in particolare neue varietà a tre dimensioni.*, Rend. Acc. Lincei, **12** (1903), pp. 409–420. (JFM **34**, p. 658.)

representation, the curve (C) is *planar*, and the curve (C') is obtained from (C) by taking through each point of (C), an infinitesimal length normal to the plane of (C). Therefore, the first variation of any arc length of (C) is zero when going from (C) to (C'). On the other hand, since the length of (C') is preserved in this representation up to second-order infinitesimals, we see that in a Riemannian manifold, the first variation of the arc length of (C) is zero when going from the plane (P) to the infinitesimally close plane (P').   □

An interesting consequence arises from this. In one of the planes (P_0) of the family, we take an arbitrary coordinate system u, v and denote by w a variable parameter defining the different planes of the family. Then, the line element of the manifold is of the form

$$ds^2 = d\sigma^2 + H(u, v, w)dw^2, \tag{21.1}$$

where $d\sigma^2$ is the line element of the plane (P_0):

$$ds^2 = E(u, v)du^2 + 2F(u, v)dudv + G(u, v)dv^2. \tag{21.2}$$

The line element (21.1) so obtained contains only one arbitrary function of three arguments (and functions of two arguments). It follows that Riemannian manifolds admitting a one-parameter family of planes are exceptional, since, as Riemann noticed, the most general line element ds^2 in three variables depends on three arbitrary functions of three variables (there are six arbitrary coefficients, but the possibility of arbitrary change in coordinates reduces the number of these functions to three).

Such a transformation must preserve the basis vector $\mathbf{I}_3$, and as a result, it must preserve dz:

$$dz = dz'.$$

This implies that

$$z' = z + c.$$

We have arrived at the conclusion of No. **127**.

E. EMBEDDED MANIFOLDS

Chapter 22

Lines in a Riemannian Manifold

132. The Frenet formulas in a Riemannian manifold. The theory of curvature of lines in a Riemannian manifold is the same as in Euclidean space.

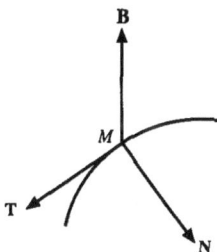

Fig. 13

We remind the reader of the Frenet formulas in Euclidean space (Fig. 13):

$$
\begin{cases}
\dfrac{d\mathbf{M}}{ds} = \mathbf{T}, \\[2mm]
\dfrac{d\mathbf{T}}{ds} = \dfrac{1}{\rho}\mathbf{N}, \\[2mm]
\dfrac{d\mathbf{N}}{ds} = -\dfrac{1}{\rho}\mathbf{T} + \dfrac{1}{\tau}\mathbf{B}, \\[2mm]
\dfrac{d\mathbf{B}}{ds} = -\dfrac{1}{\tau}\mathbf{N},
\end{cases}
$$

where $\frac{1}{\rho}$ is the curvature and $\frac{1}{\tau}$ is the torsion of the curve.

In a Riemannian manifold, first, we find (cf. No. **70**):

$$\frac{d\mathbf{M}}{ds} = \mathbf{T},$$

and second, we can write

$$\frac{d\mathbf{T}}{ds} = \frac{1}{\rho}\mathbf{N}.$$

Let us take

$$\mathbf{T} = \mathbf{I}_1, \quad \mathbf{N} = \mathbf{I}_2, \quad \mathbf{B} = \mathbf{I}_3. \tag{22.1}$$

Then

$$\omega^1 = ds, \quad \omega^2 = 0, \quad \omega^3 = 0, \tag{22.2}$$

since

$$d\mathbf{M} = \mathbf{I}_1\, ds.$$

Next,

$$\omega_1^2 = \frac{ds}{\rho}, \quad \omega_1^3 = 0,$$

since

$$D\mathbf{T} = D\mathbf{I}_1 = \frac{ds}{\rho}\mathbf{I}_2. \tag{22.3}$$

Set

$$\omega_2^3 = \frac{ds}{\tau}. \tag{22.4}$$

Then we find the remaining two equations

$$\begin{cases} \dfrac{D\mathbf{N}}{ds} = -\dfrac{1}{\rho}\mathbf{T} + \dfrac{1}{\tau}\mathbf{B}, \\[2mm] \dfrac{D\mathbf{B}}{ds} = -\dfrac{1}{\tau}\mathbf{N}. \end{cases}$$

In an n-dimensional manifold, we obtain a series of equations, in each of which a new parameter (the second, third, etc. curvatures) and a new vector (the second, third, etc. normal) are introduced.

133. Determination of a curve with given curvature and torsion. Zero torsion curves in a space of constant curvature. *Do curves exist whose curvature and torsion are given functions of the arc length?*

We attach the most general frame to a point M and set

$$\begin{cases} \omega^1 = ds, \quad \omega^2 = 0, \quad \omega^3 = 0, \\[2mm] \omega_1^2 = \dfrac{ds}{\rho}, \quad \omega_1^3 = 0, \quad \omega_2^3 = \dfrac{ds}{\tau}. \end{cases}$$

This is a system of differential equations that admits solutions defining coordinates of a point of the curve and its Frenet trihedron.

REMARK. Each curve can be developed in Euclidean space. The developed curve has the same curvature and the same torsion as the original one.

Curves with zero torsion. Let us assume that

$$\frac{1}{\tau} = 0.$$

Then

$$\frac{D\mathbf{B}}{ds} = 0.$$

Thus, the binormal remains parallel to itself at all points. In Euclidean space we have plane curves.

In a *space of constant curvature*, applying formulas (18.6) and (18.7),

$$\begin{cases} d\mathbf{M} = \omega^i\,\mathbf{I}_i, \\ d\mathbf{I}_i = \omega_i^k\,\mathbf{I}_k, \end{cases}$$

and equations (22.1), (22.2), and (22.3), we obtain

$$\begin{cases} d\mathbf{M} = \omega^1\,\mathbf{I}_1, \\ d\mathbf{I}_1 = \omega_1^2\,\mathbf{I}_2 - K\omega^1\,\mathbf{M}, \\ d\mathbf{I}_2 = \omega_2^1\,\mathbf{I}_1 + K\omega_2^3\,\mathbf{I}_3, \\ d\mathbf{I}_3 = \omega_3^2\,\mathbf{I}_2, \end{cases}$$

or

$$\begin{cases} \dfrac{d\mathbf{M}}{ds} = \mathbf{T}, \\[4pt] \dfrac{d\mathbf{T}}{ds} = \dfrac{1}{\rho}\mathbf{N} - K\mathbf{M}, \\[4pt] \dfrac{d\mathbf{N}}{ds} = -\dfrac{1}{\rho}\mathbf{T} + \dfrac{1}{\tau}\mathbf{B}, \\[4pt] \dfrac{d\mathbf{B}}{ds} = -\dfrac{1}{\tau}\mathbf{N}. \end{cases} \tag{22.5}$$

If $\dfrac{1}{\tau} = 0$, then

$$\frac{d\mathbf{B}}{ds} = 0,$$

and the analytic point $\mathbf{B}$ is fixed.

Since the scalar product $\mathbf{M} \cdot \mathbf{B}$ vanishes,

$$\mathbf{M} \cdot \mathbf{B} = 0,$$

the point $\mathbf{M}$ belongs to the polar plane of the point $\mathbf{B}$. As a result, the curve is *planar*.

For constant ρ,

$$\rho = \text{const},$$

we have

$$\frac{d}{ds}\left(\mathbf{M} + \rho\mathbf{N}\right) = \mathbf{T} + \rho\left(-\frac{1}{\rho}\mathbf{T}\right) = 0,$$

and the point

$$\mathbf{M} + \rho\mathbf{N}$$

is fixed. The scalar square of this point is

$$(\mathbf{M} + \rho\mathbf{N})^2 = \mathbf{M}^2 + 2\rho\,\mathbf{M}\cdot\mathbf{N} + \rho^2\mathbf{N}^2 = \frac{1}{K} + \rho^2.$$

If $K > 0$, then we set

$$\mathbf{C} = \frac{\mathbf{M} + \rho\mathbf{N}}{\sqrt{1 + K\rho^2}}.$$

It follows that

$$\mathbf{C}^2 = \frac{1}{K}.$$

This shows that $\mathbf{C}$ is a geometric point.

We compute the scalar product $\mathbf{M}\cdot\mathbf{C}$:

$$\mathbf{M}\cdot\mathbf{C} = \frac{\mathbf{M}^2 + \rho\,\mathbf{M}\cdot\mathbf{N}}{\sqrt{1 + K\rho^2}} = \frac{1}{K\sqrt{1 + K\rho^2}}. \tag{22.6}$$

On the other hand, applying equation (18.17) and setting $\mathbf{M}_0 = \mathbf{C}$ and the value of the arc length of the geodesic

$$s = \mathbf{M}_0\cdot\mathbf{M} = \mathbf{C}\cdot\mathbf{M} = d,$$

we obtain

$$\mathbf{M}\cdot\mathbf{C} = \left[\mathbf{C}\cos(s\sqrt{K}) + \frac{\mathbf{T}_0}{\sqrt{K}}\sin(s\sqrt{K})\right]\cdot\mathbf{C}$$
$$= \frac{1}{K}\cos(d\sqrt{K}), \tag{22.7}$$

since $\mathbf{T}_0\cdot\mathbf{C} = 0$ as a product of a vector and a point.

Comparing equations (22.6) and (22.7), we find that

$$\cos(d\sqrt{K}) = \frac{1}{\sqrt{1 + K\rho^2}}. \tag{22.8}$$

This implies that the distance d between the points C and M is constant, if K and ρ are constants.

Therefore, the locus of points M is a circle centered at C and of radius ρ. But if we find ρ from equation (22.7), we obtain

$$K\rho^2 = \tan^2(d\sqrt{2})$$

and

$$\rho = \frac{1}{\sqrt{K}} \tan(d\sqrt{2}).$$

Thus, the radius of curvature of a circle in a non-Euclidean space $(K > 0)$ *does not coincide* with its radius d (an example is: for a sphere, the radius of geodesic curvature of a parallel—a small circle—is not equal to the distance of a point from the center of the circle).

Therefore, if $K > 0$, then the curves with zero torsion and constant curvature are circles.

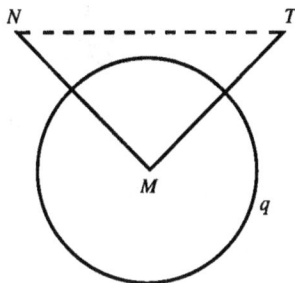

Fig. 14

134. Curves with zero torsion and constant curvature in a space of constant negative curvature. Equations (22.5) take the form

$$\begin{cases} \dfrac{d\mathbf{M}}{ds} = \mathbf{T}, \\[2mm] \dfrac{d\mathbf{T}}{ds} = -K\,\mathbf{M} + \dfrac{1}{\rho}\,\mathbf{N}, \\[2mm] \dfrac{d\mathbf{N}}{ds} = -\dfrac{1}{\rho}\,\mathbf{T}; \quad K = -\dfrac{1}{R^2}. \end{cases} \tag{22.9}$$

Consider a plane curve $(\frac{1}{\tau} = 0)$ of constant curvature $\frac{1}{\rho}$. Let q be a cross-section of the absolute by the plane of this curve, and let $\overset{*}{N}$ and T be the points representing the vectors $\mathbf{N}$ and $\mathbf{T}$ (Fig. 14). The triangle ATN is autopolar with respect to the absolute. The fixed point $\mathbf{M} + \rho\mathbf{N}$ has the scalar square equal to $\rho^2 - R^2$. We must consider three cases:

1°. $\rho < R$. The point $\mathbf{M}+\rho\mathbf{N}$ is situated inside the absolute. The corresponding geometric point is determined by the condition that its scalar square is $-R^2$.

If we set

$$\mathbf{C} = \lambda\,(\mathbf{M} + \rho\mathbf{N}),$$

then taking the scalar square, we obtain

$$-R^2 = \lambda^2\,(-R^2 + \rho^2),$$

and hence we have

$$\lambda = \frac{R}{\sqrt{R^2 - \rho^2}}, \quad \mathbf{C} = \frac{R}{\sqrt{R^2 - \rho^2}}\,(\mathbf{M} + \rho\mathbf{N}).$$

Thus, setting $r = CM$ and applying equation (18.17), we find that

$$\mathbf{C} \cdot \mathbf{M} = -R^2 \cosh\frac{r}{R}.$$

This implies that

$$\cosh\frac{r}{R} = \frac{R}{\sqrt{R^2 - \rho^2}}.$$

Therefore, for constant ρ, we have

$$r = \text{const.}$$

The locus of points M is an orthogonal trajectory of straight lines emanating from the fixed points C, i.e., it is a circumference of radius r and of constant curvature $\frac{1}{\rho}$. We have

$$\rho = R \tanh\frac{r}{R}.$$

2°. $\rho = R$. The point $\mathbf{M} + \rho\mathbf{N}$ is a point on the absolute. The normals at the points of the curve are parallel, i.e., they intersect each other at a point on the absolute.

The curve is the limit of a circumference whose radius increases infinitely.

3°. $\rho > R$. The point $\mathbf{M}+\rho\mathbf{N}$ is situated outside of the absolute. Suppose that a straight line AB is the polar line of this point, and P is the common point of AB and the normal to the curve (Fig. 15).

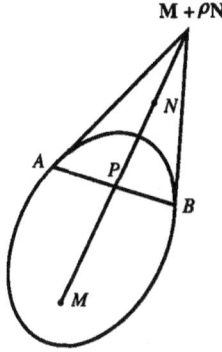

Fig. 15

The point $\mathbf{P}$ is conjugate to the point $\mathbf{M} + \rho\mathbf{N}$. It is situated on the straight line determined by the points $\mathbf{M}$ and $\mathbf{M} + \rho\mathbf{N}$. So, it can be represented as

$$\mathbf{P} = \mu\,(\mathbf{M} + \sigma\mathbf{N}) \tag{22.10}$$

with the condition

$$\mathbf{P} \cdot (\mathbf{M} + \rho\mathbf{N}) = 0.$$

Substituting the value of $\mathbf{P}$ from (22.10) into the last equation, we obtain

$$\mu\,(\mathbf{M} + \rho\mathbf{N}) \cdot (\mathbf{M} + \sigma\mathbf{N}) = \mu\,[\mathbf{M}^2 + \mathbf{M} \cdot \mathbf{N}\,(\rho + \sigma) + \rho\sigma\mathbf{N}^2] = 0,$$

and after dividing by μ, we have $-R^2 + \rho\sigma = 0$. Thus,

$$\sigma = \frac{R^2}{\rho}, \quad \mathbf{P} = \mu\left(\mathbf{M} + \frac{R^2}{\rho}\,\mathbf{N}\right).$$

Now we require that $\mathbf{P}^2 = -R^2$, i.e.,

$$\mu^2\left(\mathbf{M}^2 + \frac{R^4}{\rho^2}\,\mathbf{N}^2\right) = -R^2.$$

Applying the conditions $\mathbf{M}^2 = -R^2$ and $\mathbf{N}^2 = 1$, we see that

$$\mu^2 \left(1 - \frac{R^2}{\rho^2}\right) = 1.$$

Thus,

$$\mu = \frac{1}{\sqrt{1 - \dfrac{R^2}{\rho^2}}}, \quad \mathbf{P} = \frac{\mathbf{M} + \dfrac{R^2}{\rho}\,\mathbf{N}}{\sqrt{1 - \dfrac{R^2}{\rho^2}}}. \tag{22.11}$$

If we denote the distance PM by d, then setting $\mathbf{M}_0 = \mathbf{P}$ and $s = d$ and applying formula (18.20), we have

$$\mathbf{M} = \mathbf{P}\cosh\left(\frac{d}{R}\right) + \mathbf{T}\sinh\left(\frac{d}{R}\right).$$

Multiplying both sides of this equation by $\mathbf{P}$ and keeping in mind that $\mathbf{P}^2 = -R^2$ and $\mathbf{P} \cdot \mathbf{T} = 0$, we obtain

$$\mathbf{P} \cdot \mathbf{M} = -R^2 \cosh\left(\frac{d}{R}\right).$$

Substituting the value of $\mathbf{P}$ from formula (22.11) into the last equation, we find that

$$-\frac{R^2}{\sqrt{1 - \dfrac{R^2}{\rho^2}}} = -R^2 \cosh\left(\frac{d}{R}\right)$$

and

$$\cosh\left(\frac{d}{R}\right) = \frac{1}{\sqrt{1 - \dfrac{R^2}{\rho^2}}}.$$

It follows that for constant ρ and R, we have $d = \text{const.}$ and

$$\rho^2 = R^2 \frac{\cosh^2\left(\dfrac{d}{R}\right)}{\cosh^2\left(\dfrac{d}{R}\right) - 1} = R^2 \frac{\cosh^2\left(\dfrac{d}{R}\right)}{\sinh^2\left(\dfrac{d}{R}\right)},$$

or

$$\rho = R\coth\left(\frac{d}{R}\right).$$

The curve is the locus of the ends of segments of length d taken on perpendiculars at different points of the straight line AB. This locus is an orthogonal trajectory of perpendiculars. This is the curve parallel to the straight line AB. Therefore, if we increase the radius of curvature ρ of

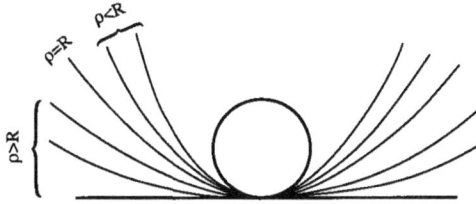

Fig. 16

the curves passing through the point M, then we obtain consecutively circumferences of increasing radii, the circumference of infinite radius and of curvature $\frac{1}{R}$, and finally curves with decreasing curvatures whose limit is a straight line (Fig. 16).

135. Integration of Frenet's equations of these curves. The Frenet formulas compose a differential system with constant coefficients, and this system can be integrated without any difficulties.

Denoting the derivatives with respect to the arc length s by primes, we have

$$\mathbf{M}''' + \left(\frac{1}{\rho^2} - \frac{1}{R^2} \right) \mathbf{M}' = 0.$$

$1°$. $\rho < R$. The integral depends on trigonometric functions of the argument

$$s\sqrt{\frac{1}{\rho^2} - \frac{1}{R^2}}.$$

Thus, the arc length L of the circumference is

$$L = \frac{2\pi}{\sqrt{\frac{1}{\rho^2} - \frac{1}{R^2}}} = \frac{2\pi\rho}{\sqrt{1 - \frac{\rho^2}{R^2}}} = 2\pi R \sinh\left(\tanh^{-1}\left(\frac{\rho}{R} \right) \right).[1]$$

$2°$. $\rho > R$. Setting

$$\frac{1}{\rho^2} - \frac{1}{R^2} = -\frac{1}{a^2},$$

we have

$$\mathbf{M} = \mathbf{A}\, e^{\frac{s}{a}} + \mathbf{B}\, e^{-\frac{s}{a}} + \mathbf{C}, \tag{22.12}$$

and hence we obtain

$$\mathbf{T} = \frac{\mathbf{A}}{a}\, e^{\frac{s}{a}} - \frac{\mathbf{B}}{a}\, e^{-\frac{s}{a}}. \tag{22.13}$$

---

[1]In the original, $2\pi R \sinh\left(\frac{\rho}{R} \right)$ was erroneously written on the right-hand side. (*Translator's note.*)

In (22.12), $\mathbf{A}, \mathbf{B}$, and $\mathbf{C}$ are three fixed analytic points defined by the conditions $\mathbf{M}^2 = -R^2$ and $\mathbf{T}^2 = 1$ for any s. This implies that

$$\mathbf{A}^2 = 0, \quad \mathbf{B}^2 = 0, \quad \mathbf{A} \cdot \mathbf{C} = 0, \quad \mathbf{B} \cdot \mathbf{C} = 0,$$

$$\mathbf{C}^2 + 2\mathbf{A} \cdot \mathbf{B} = -R^2, \quad \text{or} \quad \mathbf{A} \cdot \mathbf{B} = -\frac{a^2}{2},$$

$$1 = -\frac{2}{a^2} \mathbf{A} \cdot \mathbf{B}, \quad \mathbf{C}^2 = a^2 - R^2 > 0.$$

As a result, A and B are two points on the absolute, and C is the pole of the straight line AB (Fig. 17).

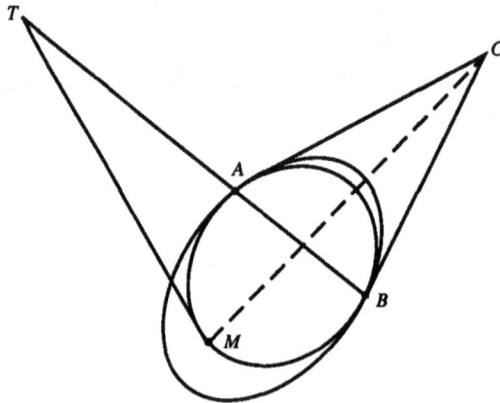

Fig. 17

Take ABC as a coordinate triangle. Then

$$\mathbf{M} = x\,\mathbf{A} + y\,\mathbf{B} + z\,\mathbf{C}.$$

Equation (22.12) shows that for any point of the curve, we have

$$x = e^{\frac{s}{a}}, \quad y = e^{-\frac{s}{a}}, \quad z = 1,$$

and as a result, the equation of the curve is

$$xy = z^2.$$

This is the curve which is tangent to the absolute twice, at the points A and B. When s changes from $-\infty$ to $+\infty$, we obtain only one of the arcs of the curve, namely the arc AMB. In the chosen coordinate system, the equation of the absolute is obtained from the condition

$$(x\,\mathbf{A} + y\,\mathbf{B} + z\,\mathbf{C})^2 = 0,$$

which implies

$$xy - \frac{a^2 - R^2}{a^2} z^2 = 0.$$

The point $\mathbf{T} = x\,\mathbf{A} + y\,\mathbf{B}$ lies on the straight line AB. Thus, the point T is the common point of AB and the tangent to the curve at the point M. It follows that MC is the normal to the curve at the point M, and N is the point on MC, which is conjugate to M with respect to the absolute. The point C is the fixed point considered above.

$3°$. $\rho = R$. This case is obtained from the previous one by passage to the limit. We find a conic hyperosculating to the absolute. The circumferences of radius $\rho < R$ are tangent to the absolute too, but the chord of tangency is external, and the tangency is imaginary.

136. Euclidean space of conjugacy. In an arbitrary Riemannian manifold, consider a curve (C), whose curvature $\frac{1}{\rho}$ and torsion $\frac{1}{\tau}$ are defined at any point M.

Attach the Frenet trihedron to every point of the curve (C). At any point M_0 homologous to the point M, the development (C_0) of the curve (C) in Euclidean space has the same curvature $\frac{1}{\rho}$ and the same torsion $\frac{1}{\tau}$ as (C). Moreover, homologous arcs have the same length.

Thus, we have a correspondence between the points M of the curve (C) and the points M_0 of the curve (C_0).

We extend this correspondence to the spaces around the curves (C) and (C_0).

To this end, we note that to a point P in a neighborhood of the curve (C), there corresponds one and only one geodesic (G), which is normal to the curve (C) at a certain point M, such that the arc length PM is a certain quantity u. The direction cosines of the curve (G) with respect to the Frenet trihedron at the point M have certain definite values $0, \beta$, and γ.

To the point P, we put in correspondence the point P_0 defined as follows. At the point M_0, we take the normal to the curve (C_0) which has the direction cosines $0, \beta$, and γ with respect to the Frenet trihedron of the curve (C_0). On this normal, we take the arc of length $M_0P_0 = u$. The Euclidean space of the points (P_0) is the *space of conjugacy* along the curve (C).

We prove that the arc $P_0Q_0 = \sigma_0$ of the space of conjugacy differs from the arc $PQ = \sigma$ of the Riemannian manifold only by second-order infinitesimals with respect to the maximum distance u. This implies that an observer who moves along the curve (C) and measures up to first-order infinitesimals could not determine the nature of the space around the curve (C).

In order to show this, we take $l = MN = M_0N_0$ (a point N is attached to the point Q) and consider a family of curves obtained by taking the arcs of length ku, $k < 1$, on the geodesics $MP, \dots, MQ$. The lines of this family depend on a parameter a. We assume that $a = 0$ for MN, $a = a_0$ for PQ, and $a = ka_0$ for the curve corresponding to the length ku. It follows that if we consider a as the amount of time that is needed for a moving point to pass the geodesic normals, then the motion of this point is uniform.

For a curve of a family, we take

$$\mathbf{I}_1 = \mathbf{T}.$$

Thus,

$$\omega^2 = 0, \quad \omega^3 = 0.$$

For $a = 0$, we have

$$
\begin{cases}
\omega_1^2 = \dfrac{ds}{\rho}, & \omega_1^3 = 0, \\[2mm]
\omega_2^3 = \dfrac{ds}{\tau}, & \omega^1 = ds.
\end{cases}
$$

If we define the symbol of differentiation

$$\delta = \frac{\partial}{\partial a},$$

then from formula (19.5) (see No. **118**), we have

$$\delta\lambda = (e^1)_N - (e^1)_M + \int_M^N (e^2\,\omega_2^1 + e^3\,\omega_3^1).$$

But e^1, the tangent component of the velocity along the line (G), is zero. Hence,

$$\delta\lambda = -\int_M^N e^2\,\frac{ds}{\rho}.$$

This proves that

$$\delta\lambda = \delta\lambda_0,$$

and this concludes our proof.

137. The curvature of a Riemannian manifold occurs only in infinitesimals of second order. The above proof does not cover all possible arcs PQ. In order to have a more complete proof, another approach is needed.

The point P is defined by the geodesic distance r from the point M and by the parameters

$$u = \beta r, \quad v = \gamma r,$$

where β and γ are the direction cosines of the geodesic normal MP. For an arbitrarily given point P, the trihedron at the point P is the Frenet trihedron at the point M, which is transported by parallelism along MP.

We compare the line elements ds^2 of the Euclidean space and the Riemannian manifold.

In *Euclidean space*, we have

$$\mathbf{P} = \mathbf{M} + u\mathbf{N} + v\mathbf{B},$$

$$d\mathbf{P} = ds\,\mathbf{T} + du\,\mathbf{N} - \tfrac{u}{\rho}\,ds\,\mathbf{T} + \tfrac{u}{\rho}\,ds\,\mathbf{B} + dv\,\mathbf{B} - \tfrac{v}{\tau}\,ds\,\mathbf{N},$$

and hence

$$\begin{cases} \varpi^1 = \left(1 - \dfrac{u}{\rho}\right)ds, \quad \varpi^2 = du - \dfrac{v}{\tau}\,ds, \quad \varpi^3 = dv + \dfrac{u}{\tau}\,ds, \\[2mm] \varpi_1^2 = \dfrac{ds}{\rho}, \qquad\qquad \varpi_1^3 = 0, \qquad\qquad \varpi_2^3 = \dfrac{ds}{\tau}. \end{cases}$$

In the *Riemannian manifold*, we have

$$\begin{cases} \omega_i^j = \varpi_i^j + \tilde{\omega}_i^j, \\[2mm] \omega_i^j = \varpi_i^j + \tilde{\omega}_i^j. \end{cases}$$

When we move along the curve MP, we have

$$ds = 0, \quad \frac{u}{v} = \text{const.},$$

or

$$u\,dv - v\,du = 0,$$

and the trihedron moves parallel to itself ($\omega_i^j = 0$). But we already have $\varpi_i^j = 0$, and hence we must have $\tilde{\omega}_i^j = 0$. Thus, ω_i^j have the form

$$\omega_i^j = \varpi_i^j + A_i^j\,ds + B_i^j\,(u\,dv - v\,du).$$

When we move along the curve MP, this curve MP slides along itself, the tangential components of a displacement are constantly zero, and the remaining components are proportional to the invariant quantities β and γ, or u and v:

$$\left.\begin{aligned} \omega^1 &= 0, \\ \omega^2 &= \beta\,dr = du, \\ \omega^3 &= \gamma\,dr = dv \end{aligned}\right\} \quad \text{along the line } MP,$$

or

$$\left.\begin{aligned} \varpi^1 &= 0, \\ \varpi^2 &= du, \\ \varpi^3 &= dv \end{aligned}\right\} \quad \text{along the line } MP.$$

Therefore, we can write

$$\omega^i = \overline{\omega}^i + A_i\, ds + B_i\, (u\, dv - v\, du).$$

If we take $u = v = 0$, then $A_i = A_i^j = 0$ (we remain on the curve (C), and as a result $\omega^i = \overline{\omega}^i$ and $\omega_i^j = \overline{\omega}_i^j$). The quantities A and B are determined by means of successive approximations. For the first-order terms, we have

$$d\omega_i^j = \frac{\partial A_i^j}{\partial u}\, du \wedge ds + \frac{\partial A_i^j}{\partial v}\, dv \wedge ds + 2B_i^j\, du \wedge dv = \omega_i^k \wedge \omega_k^j + \Omega_i^j.$$

We take $u = v = 0$. Then $\omega_i^k \wedge \omega_k^j = 0$, and we have

$$\Omega_i^j = R_{i12}^j\, ds \wedge du + R_{i13}^j\, ds \wedge dv + R_{i23}^j\, du \wedge dv.$$

Thus,

$$A_i^j = -(R_{i12}^j\, u + R_{i13}^j\, v),$$

$$B_i^j = \frac{1}{2}\, R_{i23}^j.$$

Similarly, for the first-order terms in $d\omega^i$, we have

$$d\overline{\omega}^i + \frac{\partial A_i}{\partial u}\, du \wedge ds + \frac{\partial A_i}{\partial v}\, dv \wedge ds + 2B_i\, du \wedge dv = \omega^k \wedge \omega_k^i.$$

For $u = v = 0$, the right-hand side reduces to $\overline{\omega}^k \wedge \overline{\omega}_k^i = d\overline{\omega}^i$, and

$$\frac{\partial A_i}{\partial u} = 0, \quad \frac{\partial A_i}{\partial v} = 0, \quad B_i = 0.$$

We again obtain the result stated above.

In order to find the form of the terms of first-order smallness in the expansion of $d\omega^i$, we set

$$\left\{
\begin{aligned}
A_1 &= \frac{1}{2}\left(R_{12,12}\, u^2 + 2R_{12,13}\, u\, v + R_{13,13}\, v^2\right),\\[4pt]
A_2 &= \frac{1}{2}\left(R_{23,12}\, u\, v + R_{23,13}\, v^2\right),\\[4pt]
A_3 &= \frac{1}{2}\left(R_{23,12}\, u^2 + R_{23,13}\, u\, v\right),\\[4pt]
B_1 &= -\frac{1}{6}\left(R_{23,12}\, u + R_{23,13}\, v\right),\\[4pt]
B_2 &= \frac{1}{6}\, R_{23,23}\, v,\\[4pt]
B_3 &= \frac{1}{6}\, R_{23,13}\, u.
\end{aligned}
\right.$$

It follows that

$$ds^2 = \overline{ds}^2 + (R_{12,12}\, u^2 + 2R_{12,13}\, u\, v + R_{13,13}\, v^2)\, ds^2$$

$$- \frac{4}{3} \left(R_{12,23}\, u + R_{13,23} \right) ds\, (u\, dv - v\, du) + \frac{1}{3}\, R_{23,23}\, (u\, dv - v\, du)^2.$$

It is remarkable that the additional terms do not introduce the fundamental invariants ρ and τ of the curve (C).

The above formula generalizes the formula of Riemann giving ds^2 in the normal coordinates $(\alpha r, \beta r, \gamma r)$.

Chapter 23

Surfaces in a Three-Dimensional Riemannian Manifold

138. The first two structure equations and their geometric meaning. Take the vector $\mathbf{I}_3$ as the surface normal. Then on a surface, we have

$$\omega^3 = 0.$$

The first structure equations (see (17.1)) take the form

$$\begin{cases} d\omega^1 = \omega^2 \wedge \omega_2^1, \\ d\omega^2 = \omega^1 \wedge \omega_1^2, \\ 0 = \omega^1 \wedge \omega_1^3 + \omega^2 \wedge \omega_2^3. \end{cases} \tag{23.1}$$

The geometric meaning of the first two equations of (23.1) is the following. If we consider a surface as a two-dimensional Riemannian manifold, then the forms ω_i^j of this manifold are the same as those in the ambient three-dimensional manifold.

This implies that *the geometric variation $D\mathbf{X}$ of a vector $\mathbf{X}$, which is tangent to the surface, is equal to the tangential component of the geometric variation of this vector considered as a vector of the ambient manifold.*

If the geometric variation in the ambient manifold is normal to the surface, then *the vector $\mathbf{X}$ moves along the surface parallel to itself.* Thus, we come to a new definition of parallel transport.

In particular, let us suppose that the ambient manifold is a Euclidean space. Then we construct a vector $\mathbf{X}'$ equipollent to the vector $\mathbf{X}$ in the Euclidean sense, and project it onto the tangent plane at a point M. This is Levi-Civita's point of view.

139. The third structure equation. Invariant forms (scalar and exterior). Consider now the third equation in (23.1). By Cartan's lemma, this equation implies the following linear relations:

$$\begin{cases} \omega_1^3 = a_{11}\,\omega^1 + a_{12}\,\omega^2, \\ \omega_2^3 = a_{12}\,\omega^1 + a_{22}\,\omega^2. \end{cases} \qquad (23.2)$$

The fact that the forms ω_1^3 and ω_2^3 are linear combinations of the forms ω^1 and ω^2 follows also from the derivational formula

$$D\mathbf{I}_3 = \omega_3^1\,\mathbf{I}_1 + \omega_3^2\,\mathbf{I}_2.$$

This formula shows that $D\mathbf{I}_3 = 0$, if the point M is fixed, i.e., the equations $\omega^1 = 0$, $\omega^2 = 0$ imply that $\omega_1^3 = 0$, $\omega_2^3 = 0$. But in addition, from equations (23.2) we see that the tensor (a_{ij}) is symmetric.

We conclude that (a_{ik}) is a tensor by noticing that the quadratic form

$$\varphi = \omega^1\,\omega_1^3 + \omega^2\,\omega_2^3$$

has an extrinsic meaning:

$$\varphi = -d\mathbf{M} \cdot D\mathbf{I}_3.$$

This form is the *second fundamental (Gauss) form* of the surface.

In addition to the two fundamental forms

$$\begin{cases} ds^2 = (\omega^1)^2 + (\omega^2)^2, \\ \varphi = \omega^1\,\omega_1^3 + \omega^2\,\omega_2^3, \end{cases}$$

there are two *scalar quadratic forms*,

$$d\sigma^2 = (\omega_1^3)^2 + (\omega_2^3)^2$$

—*the line element of the spherical image of the surface*, and

$$\psi = \begin{vmatrix} \omega^1 & \omega^2 \\ \omega_1^3 & \omega_2^3 \end{vmatrix}$$

—*the Jacobian of the forms φ and ds^2* (the area of the parallelogram with sides $d\mathbf{M}$ and $D\mathbf{I}_3$), and three *exterior quadratic forms*:

$$\begin{cases} \omega^1 \wedge \omega^2 \text{—the surface area element;} \\[2mm] \omega_1^3 \wedge \omega_2^3 \text{—the area element of the spherical image;} \\[2mm] \omega^1 \wedge \omega_2^3 - \omega^2 \wedge \omega_1^3. \end{cases}$$

140. The second fundamental form of a surface. We return to the form φ. In order to find its geometric meaning, we note that (ω^1, ω^2) determines the displacement ds along a certain line of the surface.

Take the tangent vector $\mathbf{T}$ to this line as the first basis vector $\mathbf{I}_1$:

$$\mathbf{T} = \mathbf{I}_1.$$

Then

$$\omega^1 = ds, \quad \omega^2 = 0,$$

and

$$\varphi = a_{11}\, ds^2.$$

We have

$$\frac{D\mathbf{T}}{ds} = \frac{D\mathbf{I}_1}{ds} = \frac{\omega_1^2\, \mathbf{I}_2 + \omega_1^3\, \mathbf{I}_3}{ds} = \frac{\omega_1^2}{ds}\mathbf{I}_2 + \frac{a_{11}\,\omega^1 + a_{12}\,\omega^2}{ds}\mathbf{I}_3,$$

i.e.,

$$\frac{D\mathbf{T}}{ds} = \frac{1}{\rho}\mathbf{N} = \frac{\omega_1^2}{ds}\mathbf{I}_2 + a_{11}\,\mathbf{I}_3.$$

Thus, a_{11} is the projection of the vector of curvature $\dfrac{1}{\rho}\mathbf{N}$ onto the surface normal. This is the *normal curvature* $\dfrac{1}{\rho_n}$. As a result,

$$\varphi = \frac{ds^2}{\rho_n}.$$

The number

$$\frac{1}{\rho_n} = \frac{\varphi}{ds^2} = \frac{\omega^1\omega_1^3 + \omega^2\omega_2^3}{(\omega^1)^2 + (\omega^2)^2}$$

depends only on the ratio $\omega^2 : \omega^1$, i.e., on the choice of tangent but not on the individually chosen curve.

This implies the following theorem.

Theorem 1. (Meusnier) *All lines on a surface tangent to the same straight line at one point have the same normal curvature.*

REMARK. It is possible to perform the computation for the general case. Denote the angle between the vectors $\mathbf{I}_3$ and $\mathbf{T}$ by θ. Then we have

$$\mathbf{T} = \mathbf{I}_1 \cos\theta + \mathbf{I}_2 \sin\theta.$$

Thus,

$$\omega^1 = \cos\theta\, ds, \quad \omega^2 = \sin\theta\, ds. \qquad (23.3)$$

The normal component of the vector $D\mathbf{T}$ multiplied by ds is

$$\frac{ds^2}{\rho_n}\mathbf{I}_3 = (\cos\theta\, \omega_1^3 + \sin\theta\, \omega_2^3)\mathbf{I}_3\, ds = (\omega_1^3\,\omega^1 + \omega_2^3\,\omega^2)\mathbf{I}_3.$$

Denote those curvatures that correspond to the direction of the axes by $\frac{1}{R_1}$ and $\frac{1}{R_2}$ and call them the *principal curvatures*. For the axes, the second fundamental form is a sum of squares [1]

$$\varphi = \frac{(\omega^1)^2}{R_1} + \frac{(\omega^2)^2}{R_2}. \tag{23.4}$$

The curves corresponding to these *principal directions* ($\omega^2 = 0$, $\omega^1 = 1$ or $\omega^1 = 0$, $\omega^2 = 1$) are said to be the *curvature lines*.

In order to find the principal directions, it suffices to express the fact that they correspond to the extrema of the normal curvature

$$\frac{\varphi}{ds^2} = \frac{a_{11}(\omega^1)^2 + 2a_{12}\omega^1\omega^2 + a_{22}(\omega^2)^2}{(\omega^1)^2 + (\omega^2)^2}.$$

Differentiating with respect to ω^1 and with respect to ω^2, after dividing by $\frac{2}{(\omega^1)^2+(\omega^2)^2}$, we obtain

$$\begin{cases} a_{11}\omega^1 + a_{12}\omega^1 - \omega^1\dfrac{\varphi}{ds^2} = 0, \\[2mm] a_{12}\omega^1 + a_{22}\omega^1 - \omega^2\dfrac{\varphi}{ds^2} = 0. \end{cases}$$

If we eliminate φ and apply formulas (23.2), we find that

$$\frac{\omega^1}{\omega_1^3} = \frac{\omega^2}{\omega_2^3}.$$

Introduce the new form

$$\begin{aligned} \psi &= \omega^1\omega_2^3 - \omega^2\omega_1^3 \\ &= a_{12}(\omega^1)^2 + (a_{22} - a_{11})\omega^1\omega^2 - a_{12}(\omega^2)^2. \end{aligned} \tag{23.5}$$

Then to find the principal directions, we arrive at the equation

$$\psi = 0.$$

The curvature lines can also be defined as follows. If we consider the surface normal at a point M of the surface, then a neighboring point M' belongs to a curvature line if the normal at the point M' intersects the normal at the point M.

Therefore, we need to require that the geometric variation of the normal belongs to the plane element defined by the two normals. Since this variation

---

[1] As is known, the second fundamental form φ can be always reduced to a sum or difference of squares

$$\varphi = a_{11}(\omega^1)^2 + a_{22}(\omega^2)^2.$$

We denote $a_{11} = \frac{1}{R_1}$ and $a_{22} = \frac{1}{R_2}$.

is perpendicular to the unit vector of the surface normal, we must require that
this variation is parallel to the displacement MM':

$$\frac{\omega^1}{\omega_1^3} = \frac{\omega^2}{\omega_2^3},$$

This justifies the above new definition of the curvature lines.

**141. Asymptotic lines. Euler's theorem. Total and mean curvature
of a surface.** An *asymptotic line* is a line whose *normal curvature is zero*, i.e.,
for which the osculating plane is tangent to the surface. The asymptotic lines
are determined by the equation

$$\varphi = 0.$$

If

$$\varphi = \frac{(\omega^1)^2}{R_1} + \frac{(\omega^2)^2}{R_2},$$

then

$$\varphi = \left(\frac{1}{R_2} - \frac{1}{R_1}\right)\omega^1\omega^2. \tag{23.6}$$

It follows that the asymptotic tangents are inclined to the principal tangents at
the same angles (i.e., the principal directions bisect the asymptotic directions).

Theorem 2. (Euler) *The sum of normal curvatures of two orthogonal directions is constant.*

Proof. This follows from the equation

$$\frac{1}{\rho_n} = \frac{\cos^2\theta}{R_1} + \frac{\sin^2\theta}{R_2},$$

which is obtained by dividing both sides of equation (23.4) by ds^2 and applying
formulas (23.3).   □

This constant is the *mean curvature*

$$\frac{1}{R_1} + \frac{1}{R_2}$$

of the surface.

In general, we have

$$\frac{1}{R_1} + \frac{1}{R_2} = a_{11} + a_{22}. \tag{23.7}$$

The *total curvature* of the surface is defined as

$$\frac{1}{R_2} \cdot \frac{1}{R_1} = a_{11}a_{22} - (a_{12})^2. \tag{23.8}$$

The mean and total curvatures are equal to the contracted tensor (a_{ij}) and the determinant of the tensor (a_{ij}), respectively.

142. Conjugate tangents. Let $\mathbf{T}$ be the vector tangent at a point M to a curve (C) of the surface, and let $\mathbf{T}'$ be the characteristic of the envelope of the tangent planes to the surface along the curve (C) (the Euclidean space). We will say that the directions $\mathbf{T}$ and $\mathbf{T}'$ are conjugate. Suppose that d is the symbol of differentiation in the direction of $\mathbf{T}$, and δ is that in the direction of $\mathbf{T}'$.

To generalize, we define the direction $\mathbf{T}'$ as the direction which is perpendicular to two infinitesimally close surface normals at infinitesimally close points M and M' of the curve (C). Being normal to $\mathbf{I}_3$, the vector $\mathbf{T}'$ is a linear combination of the forms $\omega^1(\delta)$ and $\omega^2(\delta)$. At the same time, the vector $\mathbf{T}'$ is normal to the vector

$$D\mathbf{I}_3 = \omega_3^1(d)\,\mathbf{I}_1 + \omega_3^2(d)\,\mathbf{I}_2.$$

Thus, we have

$$\omega^1(\delta)\,\omega_1^3(d) + \omega^2(\delta)\,\omega_2^3(d) = 0,$$

or

$$a_{11}\,\omega^1(d)\,\omega^1(\delta) + a_{12}\,(\omega^2(d)\,\omega^1(\delta) + \omega^1(d)\,\omega^2(\delta)) + a_{22}\,\omega^2(d)\,\omega^2(\delta) = 0.$$

Two conjugate tangents divide harmonically the directions of the asymptotic tangents.

143. Geometric meaning of the form ψ. Let us compute the form $\frac{\psi}{ds^2}$ for $\mathbf{T} = \mathbf{I}_1$. Since we now have $\omega^2 = 0$, equation (23.5) gives

$$\frac{\psi}{ds^2} = a_{12}.$$

This result is independent of the chosen curve.

Suppose that $\tilde{\omega}$ is the angle by which we should turn the principal normal vector in the positive direction around the tangent $\mathbf{T}$ to make it coincide with the surface normal (Fig. 18). Then

$$\widehat{(\mathbf{I}_3, \mathbf{B})} = \frac{\pi}{2} - \tilde{\omega}.$$

With the help of (23.2), we find that

$$\omega_3^2 = \mathbf{I}_3 \cdot D\mathbf{I}_2 = a_{12}\,\omega^1 + a_{22}\,\omega^2.$$

But

$$\begin{cases} \mathbf{I}_2 = \mathbf{N}\sin\tilde{\omega} - \mathbf{B}\cos\tilde{\omega}, \\ \mathbf{I}_3 = \mathbf{N}\cos\tilde{\omega} + \mathbf{B}\sin\tilde{\omega} \end{cases}$$

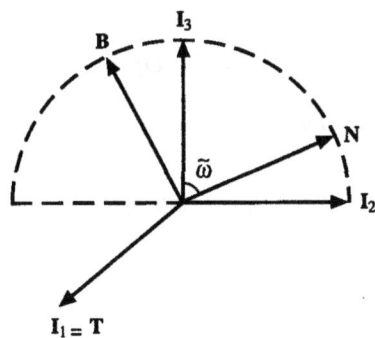

Fig. 18

and

$$\frac{D\mathbf{I}_2}{ds} = \mathbf{I}_3 \frac{\widetilde{\omega}}{ds} + \left(-\frac{1}{\rho}\mathbf{I}_1 + \frac{1}{\tau}\mathbf{B}\right)\sin\widetilde{\omega} + \mathbf{N}\frac{\cos\widetilde{\omega}}{\tau}$$

$$= \mathbf{I}_3 \frac{d\widetilde{\omega}}{ds} + \frac{1}{\tau}\left(\mathbf{N}\cos\widetilde{\omega} + \mathbf{B}\sin\widetilde{\omega}\right) - \frac{\sin\widetilde{\omega}}{\rho}\mathbf{I}_1$$

$$= \mathbf{I}_3 \left(\frac{d\widetilde{\omega}}{ds} + \frac{1}{\tau}\right) - \frac{\sin\widetilde{\omega}}{\rho}\mathbf{I}_1.$$

Thus,

$$\psi = ds^2 \left(\frac{d\widetilde{\omega}}{ds} + \frac{1}{\tau}\right)$$

The sum $\frac{d\widetilde{\omega}}{ds} + \frac{1}{\tau}$ is called the *geodesic torsion*.

Corollary 3. *All lines on a surface, having the same tangent at a common point, have the same geodesic torsion at this point.*

144. Geodesic lines on a surface. Geodesic torsion. Enneper's theorem. Geodesic lines on a surface are geodesics of a two-dimensional Riemannian manifold. They are characterized by the condition

$$D_t\mathbf{T} = 0.$$

But $D_t\mathbf{T}$ is the projection of the vector $D\mathbf{T}$ onto the tangent plane. Thus, $D\mathbf{T}$ coincides with the surface normal, or we can say that the principal normal of a geodesic is the surface normal.

The *geodesic curvature* is the curvature of a line calculated in a two-dimensional space.

Take

$$\mathbf{T} = \mathbf{I}_1.$$

Then we have

$$DI_1 = \omega_1^2\, I_2 + \omega_1^3\, I_3,$$

$$D_t I_1 = \omega_1^2\, I_2,$$

or by setting

$$\omega_1^2 = h_1\, \omega^1 + h_2\, \omega^2,$$

we obtain

$$\frac{D_t I_1}{ds} = h_1\, I_2.$$

Hence, the geodesic curvature $\dfrac{1}{\rho_g}$ is

$$\frac{1}{\rho_g} = \frac{\sin\tilde\omega}{\rho}.$$

Theorem 4 (Gauss). *The geodesic curvature depends only on the line element ds^2 of a surface.*

Proof. The theorem immediately follows from the fact that the forms ω_i^j depend on the line element only. $\square$

A geodesic corresponds to the zero geodesic curvature,

$$\frac{1}{\rho_g} = 0.$$

This implies that the angle $\tilde\omega$ between the principal normal of a geodesic and the surface normal is 0 or π.

Among all the curves tangent to each other at their common point, a geodesic line determines the normal curvature $\dfrac{1}{\rho_n}$ and the geodesic torsion $\dfrac{1}{\tau_g}$, which are the same for all these curves.

In fact, in general, we have

$$\frac{1}{\rho_n} = \frac{\cos\tilde\omega}{\rho}.$$

Therefore, for a geodesic,

$$\frac{1}{\rho_g} = \frac{1}{\rho}, \quad \frac{1}{\tau_g} = \frac{1}{\tau}.$$

This is the reason for the term *"geodesic torsion"*. It would be better to say *"normal torsion"*, since it corresponds exactly to the normal curvature.

Using expression (23.6) for the form ψ, we find that

$$\frac{1}{\tau_g} = \left(\frac{1}{R_2} - \frac{1}{R_1}\right)\sin\theta\,\cos\theta. \qquad (23.9)$$

Hence, *the curvature lines have zero geodesic torsion.*

Since for the asymptotic lines, the angle $\widetilde{\omega} = \frac{\pi}{2}$, for them the geodesic torsion coincides with the total (ordinary) torsion, as for all the lines with constant angle $\widetilde{\omega}$. We have

$$\frac{1}{\tau} = \left(\frac{1}{R_2} - \frac{1}{R_1} \right) \sin\theta \, \cos\theta$$

with the condition

$$\frac{\cos^2\theta}{R_1} + \frac{\sin^2\theta}{R_2} = 0.$$

It follows that

$$\frac{\cos^2\theta}{R_1} = -\frac{\sin^2\theta}{R_2} = \frac{1}{R_1 - R_2}.$$

Two asymptotic lines at their common point have torsions equal in absolute value and opposite in sign. The square of the torsion is

$$\frac{1}{\tau^2} = \left(\frac{R_1 - R_2}{R_1 R_2} \right)^2 \sin^2\theta \, \cos^2\theta = \frac{(R_1 - R_2)^2}{R_1 \cdot R_2} \cdot \frac{\sin^2\theta}{R_2} \cdot \frac{\cos^2\theta}{R_1} = -\frac{1}{R_1 R_2}.$$

It follows that $\frac{1}{\tau_2} = -\frac{1}{\tau_1}$. This implies that the product of the torsions $\frac{1}{\tau_1} \cdot \frac{1}{\tau_2}$ of two asymptotic lines at their common point is

$$\frac{1}{\tau_1} \cdot \frac{1}{\tau_2} = \frac{1}{R_1 R_2},$$

i.e., *the product of the torsions of two asymptotic lines at their common point is equal to the Gaussian curvature of the surface at this point* (ENNEPER'S THEOREM).

Two curves, whose tangents at their common point are mutually orthogonal, have geodesic torsions equal in absolute value and opposite in sign. In fact, by formula (23.9), the geodesic torsions of such curves differ only in sign, since

$$\sin\left(\theta + \frac{\pi}{2} \right) \cos\left(\theta + \frac{\pi}{2} \right) = -\cos\theta \, \sin\theta.$$

The same is true for any two curves whose tangents are symmetric with respect to the two principal directions. In fact, in formula (23.9), the principal directions correspond to the values $\theta = 0$ and $\theta = \frac{\pi}{2}$. The angles between any two tangents, that are symmetric with respect to the principal directions, and these principal directions differ only in sign, $\theta' = -\theta$. As a result, we again have

$$\cos\theta' \, \sin\theta' = -\sin\theta \, \cos\theta.$$

Chapter 24

Forms of Laguerre and Darboux

145. Laguerre's form. It is possible to construct other functions that, like the normal curvature $\frac{1}{\rho_n}$ and the geodesic torsion $\frac{1}{\tau_g}$, are the same for all curves tangent to one another at a point on the surface.

To this end, we consider the tensor (a_{ijk}), which is the derivative tensor of the tensor (a_{ij}). This tensor is defined by the equations

$$
\begin{cases}
Da_{11} = a_{111}\,\omega^1 + a_{112}\,\omega^2, \\[2mm]
Da_{12} = a_{121}\,\omega^1 + a_{122}\,\omega^2, \\[2mm]
Da_{22} = a_{221}\,\omega^1 + a_{222}\,\omega^2.
\end{cases}
\tag{24.1}
$$

The *Laguerre form* is defined as follows:

$$
\chi = (\omega^1)^2\,Da_{11} + 2\omega^1\,\omega^2\,Da_{12} + (\omega^2)^2\,Da_{22}.
\tag{24.2}
$$

This form is a tensor-valued form of the third degree. So, the ratio

$$
\frac{\chi}{ds^3} = \frac{a_{111}\,(\omega^1)^3 + (a_{112} + 2a_{121})\,(\omega^1)^2\,\omega^2 + (2a_{122} + a_{221})\,\omega^1\,(\omega^2)^2 + a_{222}\,(\omega^2)^3}{ds^3}
$$

admits an interpretation similar to that for the forms $\frac{\varphi}{ds^2}$ and $\frac{\psi}{ds^2}$.

Take again

$$
\mathbf{T} = \mathbf{I}_1.
\tag{24.3}
$$

Then

$$
\chi = a_{111}\,(\omega^1)^3 = a_{111}\,ds^3.
$$

But according to the general formula for absolute differentiation (see formulas (9.4) in No. **50**), we have

$$
Da_{ij} = da_{ij} - a_{kj}\,\omega_i^k - a_{ik}\,\omega_j^k,
$$

and in particular,

$$Da_{11} = da_{11} - 2a_{12}\,\omega_1^2. \tag{24.4}$$

Thus, applying (24.1) and (24.2), we obtain

$$a_{111} \equiv \frac{da_{11}}{ds} - 2\frac{\omega_1^2}{ds} \quad (\text{mod } \omega^2).$$

But for a reference system with (24.3), we have

$$\begin{cases} a_{11} = \dfrac{1}{\rho_n} = \dfrac{\cos\tilde{\omega}}{\rho}, \\[2mm] \dfrac{\omega_{12}}{ds} = \dfrac{1}{\rho_g} \equiv \dfrac{\sin\tilde{\omega}}{\rho} \quad (\text{mod } \omega^2), \\[2mm] a_{12} = \dfrac{\psi}{ds^2} = \dfrac{1}{\tau_g} = \dfrac{d\tilde{\omega}}{ds} + \dfrac{1}{\tau}. \end{cases} \tag{24.5}$$

As a result, we find that

$$\frac{\chi}{ds^3} = \frac{d}{ds}\left(\frac{\cos\tilde{\omega}}{\rho}\right) - 2\left(\frac{d\tilde{\omega}}{ds} + \frac{1}{\tau}\right)\cdot\frac{\sin\tilde{\omega}}{\rho}. \tag{24.6}$$

This formula, which holds for a reference system with (24.3), is invariant with respect to the choice of frame. Thus, it holds for any frame and for all lines with a common tangent. For geodesic lines of this family, this formula has a very simple expression

$$\frac{\chi}{ds^3} = \frac{d}{ds}\left(\frac{1}{\rho}\right). \tag{24.7}$$

146. Darboux's form. The *Darboux form* is the Jacobian of the forms ds^2 and χ:

$$\begin{vmatrix} \dfrac{\partial(ds^2)}{\partial\omega^1} & \dfrac{\partial(ds^2)}{\partial\omega^2} \\[3mm] \dfrac{\partial\chi}{\partial\omega^1} & \dfrac{\partial(\chi)}{\partial\omega^2} \end{vmatrix} = 2\omega^1\frac{\partial(\chi)}{\partial\omega^2} - 2\omega^2\frac{\partial(\chi)}{\partial\omega^1}. \tag{24.8}$$

If we apply formula (24.2), then it is easy to see that for $\omega^2 = 0$, the interpretation of the Darboux form reduces to that of the sum

$$a_{112} + 2a_{121}.$$

The tensor (a_{ijk}) is symmetric only in Euclidean space. This complicates the computation, but the method remains the same.

For an asymptotic line $\left(\tilde{\omega} = \frac{\pi}{2}\right)$, the ratio (24.6) can be written as follows:

$$-\frac{2}{\tau\rho}.$$

The Laguerre form χ contains only four coefficients out of six components of the tensor (a_{ijk}). Thus, we can construct new invariants. Such an invariant is the vector with components

$$\begin{cases} b_1 = a_{121} - a_{112}, \\[2mm] b_2 = a_{221} - a_{212}. \end{cases} \tag{24.9}$$

In fact, contracting the tensor (a_{ijk}) with the vectors X^i, Y^j, and Z^k, we obtain the invariants

$$a_{ijk} X^i Y^j Z^k \quad \text{and} \quad a_{ijk} X^i Z^j Y^k.$$

Their difference

$$a_{ijk} X^i (Y^j Z^k - Y^k Z^j)$$

is also an invariant.

But collecting the terms, for which $i, j, k = 1, 2$, we obtain

$$(Y^1 Z^2 - Y^2 Z^1)(b_1 X^1 + b_2 X^2).$$

Here the first factor is a bivector, and the second one is an invariant vector.

It is possible to write this vector in a less abstract form.

Consider a small cycle (C) on a surface and the integral along this contour,

$$-\int_C D\mathbf{I}_3 = \int_C (\omega_1^3 \mathbf{I}_1 + \omega_2^3 \mathbf{I}_2) = \int_C (a_{11} \omega^1 + a_{12} \omega^2) \mathbf{I}_1 + \int_C (a_{12} \omega^1 + a_{22} \omega^2) \mathbf{I}_2.$$

Applying exterior differentiation to the forms in the integrands, we obtain

$$d\mathbf{I}_1 \wedge (a_{11} \omega^1 + a_{12} \omega^2) + \mathbf{I}_1 (da_{11} \wedge \omega^1 + da_{12} \wedge \omega^2 + a_{11} d\omega^1 + a_{12} d\omega^2)$$

$$+ d\mathbf{I}_2 \wedge (a_{12} \omega^1 + a_{22} \omega^2) + \mathbf{I}_2 (da_{12} \wedge \omega^1 + da_{22} \wedge \omega^2 + a_{12} d\omega^1 + a_{22} d\omega^2).$$

If we note that formulas (24.1) and (24.4) give

$$\begin{cases} da_{11} = 2a_{12}\omega_1^2 + a_{111}\omega^1 + a_{112}\omega^2, \\[2mm] da_{12} = (a_{22} - a_{11})\omega_1^2 + a_{121}\omega^1 + a_{122}\omega^2, \\[2mm] da_{22} = -2a_{12}\omega_1^2 + a_{221}\omega^1 + a_{222}\omega^2, \end{cases}$$

we obtain from the previous equation (after collecting similar terms)

$$\mathbf{I}_1 \left(a_{121} - a_{112} \right) \omega^1 \wedge \omega^2 + \mathbf{I}_2 \left(a_{221} - a_{122} \right) \omega^1 \wedge \omega^2.$$

Finally, we have

$$-\int_C D\mathbf{I}_3 = \iint_S \left(b^1 \mathbf{I}_1 + b^2 \mathbf{I}_2 \right) \omega^1 \wedge \omega^2, \qquad (24.10)$$

where S is the area of the surface bounded by the cycle (C).

147. Riemannian curvature of the ambient manifold. Now we have arrived at that place in the development of the theory of surfaces in the Riemannian manifold, where it is necessary to introduce the curvature of the ambient manifold. For this purpose, to the three structure equations (23.1), we need to add three more structure equations:

$$d\omega_1^2 = \omega_1^3 \wedge \omega_3^2 + R_{12,12}\, \omega^1 \wedge \omega^2, \qquad (24.11)$$

$$\begin{cases} d\omega_1^3 = \omega_1^2 \wedge \omega_2^3 + R_{13,12}\, \omega^1 \wedge \omega^2, \\[2mm] d\omega_2^3 = \omega_2^1 \wedge \omega_1^3 + R_{23,12}\, \omega^1 \wedge \omega^2. \end{cases} \qquad (24.12)$$

The left-hand side of equation (24.11) determines the Riemannian curvature of the surface considered as a two-dimensional Riemannian manifold. We write

$$d\omega_1^2 = -K_i\, \omega^1 \wedge \omega^2 \qquad (24.13)$$

and call the scalar K_i the *internal Riemannian curvature* of the surface.

In contrast to the internal curvature, we have the *external curvature* K_e,

$$R_{12,12} = -K_e. \qquad (24.14)$$

This is the curvature of the Riemannian manifold in the direction of the tangent plane to the surface. Finally, the middle term of equation (24.11) produces

$$\omega_1^3 \wedge \omega_3^2 = -\left(a_{11}\, \omega^1 + a_{12}\, \omega^2 \right) \wedge \left(a_{12}\, \omega^1 + a_{22}\, \omega^2 \right)$$

$$= -\left(a_{11}\, a_{22} - (a_{12})^2 \right) \omega^1 \wedge \omega^2 = -\frac{1}{R_1 R_2}\, \omega^1 \wedge \omega^2.$$

Here $\frac{1}{R_1 R_2}$, the product of the principal curvatures, is called the *Euler curvature*. Thus, equation (24.11) takes the form

$$-K_i\, \omega^1 \wedge \omega^2 = -\frac{1}{R_1 R_2}\, \omega^1 \wedge \omega^2 + R_{12,12}\, \omega^1 \wedge \omega^2.$$

This equation and (24.14) imply that

$$K_i = \frac{1}{R_1 R_2} + K_e. \qquad (24.15)$$

The internal curvature of the surface is the sum of its external and Euler curvatures.

REMARK 1. Consider an observer who does not leave the surface. He knows the internal curvature of the surface, since he knows the form ω_1^2 determining the rotation of the frame around the surface normal. He knows the Euler curvature, since he can measure the maximal and minimal curvatures of the normal section of the surface. As a result, he can compute the external curvature

$$K_e = -R_{12,12},$$

i.e., one component of the curvature of the ambient manifold is known to him.

REMARK 2. Historically, everything was different. The surfaces that were investigated were those embedded into the Euclidean space, i.e., those surfaces, for which

$$K_e = 0.$$

Hence, the Gauss theorem arises: *the total curvature $\frac{1}{R_1 R_2}$ depends only on the line element ds^2 of the surface.*

148. The second group of structure equations. We now turn to equations (24.12).

The tensor with components (ω_1^3, ω_2^3) is a vector. Its absolute differential has the components

$$R_{13,12}\, \omega^1 \wedge \omega^2 \quad \text{and} \quad R_{23,12}\, \omega^1 \wedge \omega^2.$$

Therefore,

$$\int_C (\omega_1^3\, \mathbf{I}_1 + \omega_2^3\, \mathbf{I}_2) = \iint_S (R_{13,12}\, \mathbf{I}_1 + R_{23,12}\, \mathbf{I}_2)\, \omega^1 \wedge \omega^2.$$

Comparing this equation with equation (24.10), we obtain

$$b_1 = R_{13,12}, \quad b_2 = R_{23,12}. \qquad (24.16)$$

Thus, an observer situated on the surface can measure not only the component $R_{12,12}$ of the Riemannian curvature of the ambient manifold, but also two more components

$$R_{13,12} \quad \text{and} \quad R_{23,12},$$

i.e., a total of three of the six components of the curvature tensor.

The rotation associated with a cycle on the surface is defined by the components

$$p = -b_2, \quad q = b_1, \quad r = -R_{12,12}.$$

This depends only on quantities available to the observer on the surface. The vector (b_1, b_2) is perpendicular to the vector (p, q). We have $r = K_i - \frac{1}{R_1 R_2}$.

149. Generalization of classical theorems on normal curvature and geodesic torsion. We saw that the classical theorems concerning the normal curvature and the geodesic torsion of two curves tangent to one another at a surface point—the theorems arising essentially from the tensor character of the coefficients a_{ij} of the second fundamental form of the surface—admit generalizations. In No. **145** and No. **146**, we considered the first derivative tensor (a_{ijk}) defined by equations (24.1) and (24.4),

$$
\begin{cases}
Da_{11} = da_{11} - 2a_{12}\,\omega_1^2 = a_{111}\,\omega^1 + a_{112}\,\omega^2, \\[2mm]
Da_{12} = Da_{21} = da_{12} + (a_{11} - a_{22})\,\omega_1^2 \\[1mm]
\qquad = a_{121}\,\omega^1 + a_{122}\,\omega^2 = a_{211}\,\omega^1 + a_{212}\,\omega^2, \\[2mm]
Da_{22} = da_{22} + 2a_{12}\,\omega_1^2 = a_{221}\,\omega^1 + a_{222}\,\omega^2.
\end{cases}
\tag{24.17}
$$

Formula (24.6), considered for $\omega^2 = 0$, provides a geometric meaning for the component

$$
a_{111} = \frac{d\left(\dfrac{1}{\rho_n}\right)}{ds} - \frac{2}{\tau_g} \cdot \frac{1}{\rho_g}.
\tag{24.18}
$$

With the help of (24.5), the second equation of (24.17) gives

$$
a_{121} = a_{211} = \frac{da_{12}}{ds} + (a_{11} - a_{22})\frac{\omega_{12}}{ds} = \frac{d\left(\dfrac{1}{\tau_g}\right)}{ds} + \left(\frac{2}{\rho_n} - L\right)\frac{1}{\rho_g},
\tag{24.19}
$$

where

$$
L = a_{11} + a_{22} = \frac{1}{R_1} + \frac{1}{R_2}
\tag{24.20}
$$

is the mean curvature of the surface.

Finally, equations (24.9) and (24.16) give

$$
\begin{cases}
a_{112} - a_{121} = K_{13}, \\[1mm]
a_{212} - a_{221} = K_{23},
\end{cases}
$$

where, for brevity, we used the notations

$$
K_{13} = -R_{13,12} \quad \text{and} \quad K_{23} = -R_{23,12}.
$$

Similarly, the second derivative tensor of the tensor (a_{ij}) is defined by the equations

$$
\begin{cases}
Da_{111} = da_{111} - (a_{211} + a_{121} + a_{112})\,\omega_1^2 = a_{1111}\,\omega^1 + a_{1112}\,\omega^2, \\
Da_{121} = da_{121} - (a_{122} + a_{221} - a_{111})\,\omega_1^2 = a_{1211}\,\omega^1 + a_{1212}\,\omega^2.
\end{cases}
$$

For the curve $\omega^1 = ds$, $\omega^2 = 0$, this implies that

$$
\begin{cases}
a_{1111} = \dfrac{da_{111}}{ds} - \left(3a_{121} + K_{13}\right)\dfrac{1}{\rho_g}, \\[2mm]
a_{1211} = \dfrac{da_{121}}{ds} - \left(2a_{221} - a_{111} + K_{23}\right)\dfrac{1}{\rho_g}.
\end{cases}
$$

The sum of the first and last equations of (24.17) is

$$
d(a_{11} + a_{22}) = (a_{111} + a_{221})\,\omega^1 + (a_{112} + a_{222})\,\omega^2.
$$

It follows that

$$
a_{221} = \frac{d}{ds}\,(a_{11} + a_{22}) - a_{111} = \frac{dL}{ds} - a_{111}.
$$

This and equations (24.18) and (24.19) allow us to find that for the curve $\omega^1 = ds$, $\omega^2 = 0$, we have

$$
\begin{aligned}
a_{1111} = {}& \frac{d^2\left(\frac{1}{\rho_n}\right)}{ds^2} - \frac{2}{\tau_g}\cdot\frac{d\left(\frac{1}{\rho_g}\right)}{ds} - \frac{1}{\rho_g}\left[5\,\frac{d\left(\frac{1}{\tau_g}\right)}{ds} + K_{13}\right] \\
& - \frac{3}{(\rho_g)^2}\left(\frac{2}{\rho_n} - L\right),
\end{aligned}
\tag{24.21}
$$

$$
\begin{aligned}
a_{1211} = {}& \frac{d^2\left(\frac{1}{\tau_g}\right)}{ds^2} + \left(\frac{2}{\rho_n} - L\right)\cdot\frac{d\left(\frac{1}{\rho_g}\right)}{ds} \\
& + \frac{1}{\rho_g}\left[5\,\frac{d\left(\frac{1}{\rho_n}\right)}{ds} - \frac{6}{\tau_g}\cdot\frac{1}{\rho_g} - 3\,\frac{dL}{ds} - K_{23}\right].
\end{aligned}
\tag{24.22}
$$

We have thus proved the following result.

Theorem 1. *Consider different curves (C) with a common tangent at a point M on a surface in a three-dimensional Riemannian manifold. At the point M, all these curves have the same normal curvature, geodesic torsion, and four more quantities (24.18), (24.19), (24.20), and (24.22).*

Here the quantities $\frac{1}{\rho_n}, \frac{1}{\rho_g}$, and $\frac{1}{\tau_g}$ are the normal and geodesic curvature and the geodesic torsion; L is the mean curvature (the sum of principal curvatures); K_{23} is the mixed Riemannian curvature at the point M of the ambient manifold for two plane elements tangent to the curves C, one of which is tangent to the surface, and the second of which is normal to it; and finally, K_{13} is the mixed Riemannian curvature for two plane elements, one of which is tangent to the surface, and the second of which is normal to the curves (C).

The quantities K_{13} and K_{23} vanish if the Riemannian manifold is a space of constant curvature or if the surface normal coincides with the principal direction of the ambient manifold.

In Euclidean space, we have $b_1 = b_2 = 0$, and the tensor (a_{ijk}) is symmetric. The same conclusion can be made for a space of constant curvature.

150. Surfaces with a given line element in Euclidean space. *Find all surfaces with the same line element ds^2 in Euclidean space.*

We know the forms ω^1 and ω^2 obtained in a representation of the form ds^2 as a sum of squares. By means of equations (23.1), this allows us to find the form ω_1^2 and the inner curvature K_i ($K_e = 0$ in the Euclidean space). In this two-dimensional Riemannian manifold, we need to construct the tensor (a_{ij}) admissible for the given line element. This tensor must satisfy conditions (24.11) and (24.9), namely,

$$
\begin{cases}
a_{11}\,a_{22} - (a_{12})^2 = K_i, \\
a_{121} - a_{112} = 0, \\
a_{221} - a_{212} = 0.
\end{cases}
$$

There are three conditions here with three unknowns a_{11}, a_{12}, and a_{22}. One of these conditions is a condition for these components themselves, and the other two are first-order partial differential equations.

Conversely, suppose that we know a solution of this system. Attach the most general orthonormal trihedron to each point. This trihedron is defined by six parameters u_i with components of infinitesimal displacement $\overline{\omega}^i$ and $\overline{\omega}_i^j$ as functions of the parameters u_i.

Consider the system of Pfaffian equations:

$$
\begin{cases}
\overline{\omega}^1 = \omega^1, \quad \overline{\omega}_1^2 = \omega_1^2, \\
\overline{\omega}^2 = \omega^2, \quad \overline{\omega}_1^3 = \omega_1^3, \\
\overline{\omega}^3 = 0, \quad \overline{\omega}_2^3 = \omega_2^3,
\end{cases}
\tag{24.23}
$$

where the forms ω_1^3 and ω_2^3 are linear combinations of the forms ω^1 and ω^2 with coefficients a_{ij} taken from the given solution.

The system of equations (24.23) is completely integrable. In fact, since the forms ω^i and ω_i^j as well as the forms $\overline{\omega}^i$ and $\overline{\omega}_i^j$ satisfy the same structure equations of the Euclidean space, the equations obtained by exterior differentiation of (24.23) are algebraic consequences of system (24.23) itself.

The considerations are the same in a space of constant curvature.

Equations (24.12) are the *Peterson–Codazzi equations*.[1]

151. Problems on Laguerre's form. PROBLEM 1. *Do curves exist satisfying simultaneously the two equations*

$$\varphi = 0 \text{ and } \chi = 0 \ ?$$

Since for the curves in question, the second fundamental form vanishes, they must be asymptotic. Thus, we have

$$\widetilde{\omega} = \frac{\pi}{2},$$

and from formula (24.6), we find that

$$\frac{\chi}{ds^2} = -\frac{2}{\rho\tau}.$$

Two cases are possible:

1°. $\frac{1}{\rho} = 0$. The curvature is zero. An asymptotic line of the surface is a geodesic line of the ambient manifold ("the straight line").

2°. $\frac{1}{\tau} = 0$. An asymptotic line has zero curvature. By Enneper's theorem (No. **144**), the total curvature K_i of the surface is zero. Thus, an asymptotic line is a double line.

If this is the case for a family of curves (the forms φ and χ have a common factor), then in case 1°, the surface is ruled, and in case 2°, the surface has only one family of asymptotic lines (a developable surface).

Consider the latter case. Take the tangent vector of the asymptotic line as the vector $\mathbf{I}_1$. Since now this line is also a curvature line, we have

$$a_{11} = 0, \quad a_{12} = 0, \quad a_{111} = 0. \tag{24.24}$$

But formulas (24.17) give

$$da_{11} + 2a_{12}\,\omega_2^1 = a_{111}\,\omega^1 + a_{112}\,\omega^2 \tag{24.25}$$

and

$$da_{12} + (a_{22} - a_{11})\,\omega_2^1 = a_{121}\,\omega^1 + a_{122}\,\omega^2. \tag{24.26}$$

---

[1] In modern texts, these equations are known as *Mainardi–Codazzi equations*, or *Peterson–Mainardi–Codazzi equations*; see for example, the book A. Gray *Modern differential geometry of curves and surfaces with MATHEMATICA*, 2nd. ed., 1998, CRC Press LLC, Boca Raton. (MR 2000i:53001; Zbl. 942.53001.) (*Translator's note.*)

Substituting (24.24) into equation (24.25), we obtain

$$a_{112} = 0.$$

If we are in a space of constant curvature, then

$$a_{121} = a_{112} = 0.$$

Thus, along the asymptotic line $\omega^2 = 0$, equations (24.26) become

$$a_{22}\, \omega_2^1 = 0.$$

Under our conditions, the component a_{22} cannot be zero, unless the surface degenerates into a plane. So, along the asymptotic line, we have

$$\omega_2^1 = 0,$$

i.e., the geodesic curvature is zero. But the normal curvature $\frac{1}{\rho_n}$ is already zero. Hence, the curvature $\frac{1}{\rho}$ is zero, and the curve is geodesic. Therefore, in a space of constant curvature, any developable surface is ruled. This conclusion does not hold in the general case, since it is possible to find nonruled developable surfaces.

PROBLEM 2. *Do surfaces exist for which the Laguerre form $\chi = 0$?*

We note only one property of such surfaces: all their geodesics are curves of constant curvature.

PROBLEM 3. *Find surfaces, for which the tensor (a_{ijk}) is identically equal to zero, i.e., surfaces for which the Euler curvature is preserved under (tangent) parallel transport along the surface.*

The surface normal is the principal direction of the ambient manifold, since we now have

$$|(b_1, b_2)| = |(p, q)| = 0.$$

Return to equations (24.17):

$$da_{11} + 2a_{12}\, \omega_2^1 = 0, \tag{24.27}$$

$$da_{12} + (a_{22} - a_{11})\, \omega_2^1 = 0, \tag{24.28}$$

$$da_{22} + 2a_{12}\, \omega_2^1 = 0. \tag{24.29}$$

We can assume that $a_{12} = 0$. This means that the second fundamental form φ reduces to an algebraic sum of squares. Then equations (24.27) and (24.29) show that the principal curvatures a_{11} and a_{22} are constant. Equation (24.28) takes the form

$$(a_{22} - a_{11})\, \omega_2^1 = 0.$$

It follows that either $a_{11} = a_{22}$, i.e., the principal curvatures are equal, and the surface has only umbilical points $\left(\frac{1}{R_1} = \frac{1}{R_2} \right)$, or $\omega_2^1 = 0$, and

$$ds^2 = dx^2 + dy^2,$$

i.e., the surface is applicable onto a plane (the line element of a plane).

152. Invariance of normal curvature under parallel transport of a vector. An arbitrary curve tangent to a unit vector (X^1, X^2) has normal curvature

$$a_{ij} \, X^i \, X^j.$$

Transport this vector by parallelism along the surface to the unit vector $(X^{1'}, X^{2'})$. Require that the normal curvature does not change:

$$d(a_{ij} \, X^i \, X^j) = 0.$$

It follows that

$$D_t a_{ij} \cdot X^i \, X^j = 0,$$

since, by the construction, we have

$$D_t X^i = 0.$$

Thus,

$$a_{ijk} = 0.$$

We note a few consequences of this result.

1. *Parallel transport preserves the principal directions, the principal curvature, the asymptotic tangents, and the geodesic curvatures.*

2. Let M and M' be two neighboring points of a curvature line. Under parallel transport, the principal directions at the point M are transferred into the principal directions at the point M'. Therefore, *the curvature lines are geodesic lines.* The latter have zero torsion and constant curvature.

3. Suppose that there exists an *asymptotic line. This line is simultaneously a geodesic line* that is a geodesic of the ambient manifold ("the straight line").

4. Suppose that through a point M, a curvature line (C) and a geodesic line

(Γ) pass (Fig. 19).

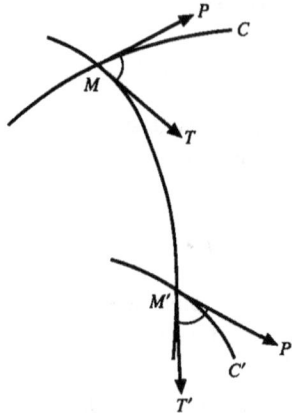

Fig. 19

If we move along the geodesic (Γ), then the tangent MT to (Γ) preserves its direction and becomes $M'T'$. Similarly, the tangent MP of the principal direction is transferred to the tangent $M'P'$ of the principal direction. Thus, *geodesic lines intersect the curvature lines at a constant angle.*

5. *Geodesics are lines of constant curvature and constant torsion.*

REMARK. All previous considerations are invalid in the case $\frac{1}{R_1} = \frac{1}{R_2}$ (indeterminate curvature lines).

In this case, the geodesic curvature is identically equal to zero. The geodesics are lines with zero torsion and constant curvature.

6. Suppose that

$$\frac{1}{R_1} = \frac{1}{R_2}.$$

Then *the Riemannian curvature of the surface is zero.*

GEOMETRIC PROOF. Consider a parallelogram $ABCD$ formed by four curvature lines (Fig. 20).

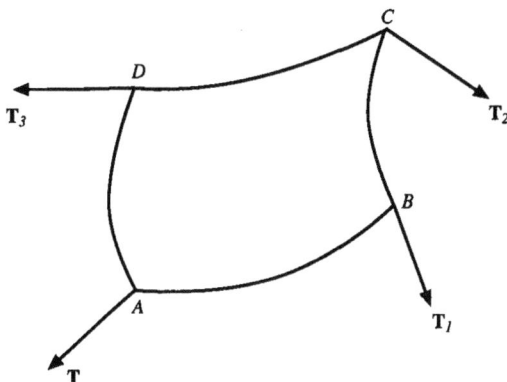

Fig. 20

Transport the tangent **T** at the point A by parallelism along $ABCD$. We return to the original vector **T**. Thus, the rotation associated with a cycle is zero, and the Riemannian curvature of the surface is zero.

153. Surfaces in a space of constant curvature. We now turn to a space of constant curvature.

First, suppose that a surface has two families of real asymptotic lines. Then the surface bears two families of rectilinear generators, i.e., it is a surface of second order.

If there is only one family of asymptotic lines, then we have developable surfaces, not necessarily of second order.

In the general case, when $\frac{1}{R_1} = \frac{1}{R_2} = 0$, the normal curvature and the geodesic torsion are identically equal to zero. The asymptotic lines are indeterminate. The geodesic lines are straight lines. If a straight line has two points belonging to such a surface, then it lies on the surface entirely (a "plane"): in general, such surfaces do not exist. The torsion of all lines on the surface is zero. Such surfaces (No. **129**) are called "*totally geodesic*". The tensor (a_{ij}) is identically equal to zero. These surfaces are normal to the principal directions of the ambient manifold.

The condition $a_{ijk} = 0$ is equivalent to the requirement

$$\frac{1}{R_1} = \text{const.}, \quad \frac{1}{R_2} = \text{const.}$$

Consider the structure equations

$$\begin{cases} d\omega^1 = \omega^2 \wedge \omega_2^1, \\ d\omega^2 = \omega^1 \wedge \omega_1^2, \\ d\omega_1^2 = \omega_1^3 \wedge \omega_3^2 - K\omega^1 \wedge \omega^2, \end{cases} \tag{24.30}$$

$$\begin{cases} d\omega_1^3 = \omega_1^2 \wedge \omega_2^3, \\ d\omega_2^3 = \omega_2^1 \wedge \omega_1^3. \end{cases} \tag{24.31}$$

Denoting the principal curvatures by $\frac{1}{\rho_1}$ and $\frac{1}{\rho_2}$, we set

$$\omega_1^3 = \frac{\omega^1}{\rho_1}, \quad \omega_2^3 = \frac{\omega^2}{\rho_2}.$$

Substituting these forms into the first equation of (24.31), we have

$$\frac{1}{\rho_1} \omega^2 \wedge \omega_2^1 = \frac{1}{\rho_2} \omega_1^2 \wedge \omega^2.$$

Therefore,

$$\left(\frac{1}{\rho_1} - \frac{1}{\rho_2} \right) \omega^2 \wedge \omega_2^1 = 0.$$

Similarly, the second equation of (24.31) gives

$$\left(\frac{1}{\rho_1} - \frac{1}{\rho_2} \right) \omega^1 \wedge \omega_1^2 = 0.$$

1°. If $\frac{1}{\rho_1} = \frac{1}{\rho_2} = \frac{1}{\rho}$, then we need to find the forms ω^1, ω^2, and ω_1^2 in such a way that

$$\begin{cases} d\omega^1 = \omega^2 \wedge \omega_2^1, \\ d\omega^2 = \omega^1 \wedge \omega_1^2, \\ d\omega_1^2 = -\left(\frac{1}{\rho^2} + K \right) \omega^1 \wedge \omega^2. \end{cases}$$

This is a two-dimensional Riemannian manifold with constant Riemannian curvature. Its existence was proved earlier.

We have

$$\begin{cases} d\mathbf{M} = \omega^1 \mathbf{I}_1 + \omega^2 \mathbf{I}_2, \\ d\mathbf{I}_1 = -K\omega^1 \mathbf{M} + \omega_1^2 \mathbf{I}_2 + \frac{1}{\rho} \omega^1 \mathbf{I}_3, \\ d\mathbf{I}_2 = -K\omega^2 \mathbf{M} + \omega_2^1 \mathbf{I}_1 + \frac{1}{\rho} \omega^2 \mathbf{I}_3, \\ d\mathbf{I}_3 = -\frac{1}{\rho} \omega^1 \mathbf{I}_1 - \frac{1}{\rho} \omega^2 \mathbf{I}_2. \end{cases} \tag{24.32}$$

The point $\mathbf{M} + \rho \mathbf{I}_3$ is fixed.

In a Euclidean or an elliptic space, these surfaces are spheres.

In a hyperbolic space, we have three classes of surfaces: spheres, quadrics twice tangent to the absolute, and quadrics having a second-order tangency with the absolute.

The Riemannian curvature of these last surfaces (horospheres) is zero $\left(\text{since } \frac{1}{\rho} = \frac{1}{R}\right)$. The geometry of the Euclidean plane is realized perfectly here. The geodesics are the plane curves with zero torsion. Their plane containing the normal $\mathbf{I}_3$ passes through a fixed point C (the point of tangency with the absolute). Thus, the geodesics are the cross-sections of a horosphere by the planes passing through the point C. These straight lines meet each other at infinity. The parallels are cross-sections by means of planes passing through the tangent to the absolute at the point C.

$2°$. If $\frac{1}{\rho_1} \neq \frac{1}{\rho_2}$, then

$$\omega^1 \wedge \omega_1^2 = 0, \quad \omega^2 \wedge \omega_2^1 = 0.$$

It follows that

$$\omega_1^2 = 0 \text{ and } \rho_1\rho_2 = -\frac{1}{K},$$

$$d\omega^1 = 0, \quad d\omega^2 = 0.$$

There are an infinite number of such surfaces.

In Euclidean space, these surfaces are circular cylinders. In the manifolds where $K \neq 0$ such a surface has nonvanishing total curvature, but its Riemannian curvature is zero.

We have

$$\begin{cases} d\mathbf{M} = \omega^1\,\mathbf{I}_1 + \omega^2\,\mathbf{I}_2, \\[2mm] d\mathbf{I}_1 = -K\,\omega^1\,\mathbf{M} + \dfrac{1}{\rho_1}\,\omega^1\,\mathbf{I}_3, \\[2mm] d\mathbf{I}_2 = -K\,\omega^2\,\mathbf{M} + \dfrac{1}{\rho_2}\,\omega^2\,\mathbf{I}_3, \\[2mm] d\mathbf{I}_3 = -\dfrac{1}{\rho_1}\,\omega^1\,\mathbf{I}_1 - \dfrac{1}{\rho_2}\,\omega^2\,\mathbf{I}_2. \end{cases}$$

It is always possible to assume that there exists a family of curvature lines formed by circumferences centered at

$$\mathbf{C} = \mathbf{M} + \rho_1\,\mathbf{I}_3$$

(it is assumed that $\rho_1 < R$). The differential of this point is

$$d\mathbf{C} = d\mathbf{M} + \rho_1\,d\mathbf{I}_3 = \omega^2\left(1 - \frac{\rho_1}{\rho_2}\right)\mathbf{I}_2. \tag{24.33}$$

The circumference is described by the point M, when the center C is fixed, i.e., when $\omega^2 = 0$. The tangent is determined by the vector $\mathbf{I}_1$. Since the plane of the circumference passes through the center, it contains two vectors $\mathbf{I}_1$ and $\mathbf{I}_3$.

The point C moves orthogonally to this plane in the direction of the vector $\mathbf{I}_2$.

Since for $K = -\frac{1}{\rho_1 \rho_2}$, we have

$$d\mathbf{I}_2 = -K\,\omega^2\,\mathbf{M} + \frac{1}{\rho_2}\,\omega^2\,\mathbf{I}_3 = -K\,\omega^2\,(\mathbf{M} + \rho_1\,\mathbf{I}_3) = -K\,\omega^2\,\mathbf{C}, \quad (24.34)$$

it follows from equations (24.33) and (24.34) that the straight line $(\mathbf{C}, \mathbf{I}_3)$ is fixed. This is similar to the Euclidean cylinder.

Consider the center (real or virtual)

$$\mathbf{M} + \rho_2\,\mathbf{I}_3$$

of the curvature lines of the second family.

Since for $\omega^2 = 0$, we have

$$d\left(\mathbf{M} + \rho_2\,\mathbf{I}_3\right) = \omega^1\,\mathbf{I}_1 - \frac{\rho_2}{\rho_1}\,\omega^1\,\mathbf{I}_1 = \omega^1\left(1 - \frac{\rho_2}{\rho_1}\right)\mathbf{I}_1,$$

this center describes the straight line in the direction $\mathbf{I}_1$, which is conjugate to $(\mathbf{C}, \mathbf{I}_3)$ with respect to the absolute.

In hyperbolic space, such a surface also produces the plane Euclidean geometry, but not perfectly, like the Euclidean cylinder.

In elliptic space, such a surface is a ruled hyperboloid. It can be considered as a cylinder of revolution around two orthogonal axes Δ and Δ'.

If we place Δ at infinity, we obtain (non-Euclidean) parallels with a constant radius. The generating hyperbolas are circumferences whose centers are points at infinity.

It is possible to consider this surface as a torus in two different ways. Here also plane Euclidean geometry is produced in the form topologically equivalent to a torus (Clifford).

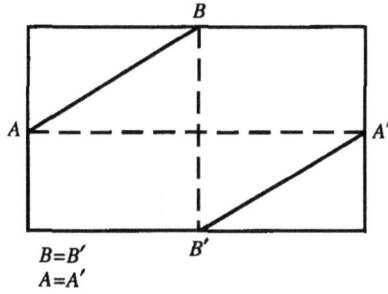

Fig. 21

In spherical space, such a surface can be represented as a rectangle with identified opposite sides (Fig. 21):

$$B \longleftrightarrow B', \quad A \longleftrightarrow A'.$$

In elliptic space it can be represented as a rhombus.

Chapter 25

p-Dimensional Submanifolds in a Riemannian Manifold of n Dimensions

154. Absolute variation of a tangent vector. Interior differentiation. Euler's curvature. At a point P of a submanifold V, we define p tangent vectors and components of a tangent displacement

$$\omega^1, \omega^2, \ldots, \omega^p,$$

and $n - p$ normal vectors with normal displacements

$$\omega^\alpha, \quad \alpha = p+1, \ldots, n.$$

Thus, the equations of V are

$$\omega^\alpha = 0, \quad \alpha = p+1, \ldots, n.$$

The first structure equations of the submanifold V are

$$dw^i = \omega^k \wedge \omega_k^i, \tag{25.1}$$

$$0 = \omega^k \wedge \omega_k^\alpha; \tag{25.2}$$

here the Latin indices take the values $1, 2, \ldots, p$, and the Greek indices take the values $p+1, \ldots, n$. Applying Cartan's lemma to equations (25.2), we obtain

$$\omega_i^\alpha = \gamma_{ik}{}^\alpha \omega^k, \quad \gamma_{ik}{}^\alpha = \gamma_{ki}{}^\alpha. \tag{25.3}$$

The absolute variation of a vector X^i tangent to the submanifold V, contains the tangent component

$$dX^i + \omega_k^i X^k$$

and the normal component

$$\omega_k^\alpha X^k.$$

The tangent component composes the absolute differential of the vector considered as a vector of the p-dimensional Riemannian manifold. The forms ω_i^j define the *interior differentiation tangent to the submanifold V*.

As for the normal component, it introduces only the value of the vector at the point P, not at neighboring points, and depends on the forms ω_k^α only. We shall say that the forms ω_k^α define the "*Euler curvature*" of the submanifold V (cf. the forms ω_1^3 and ω_2^3 in No. **139**). We will show that there also exists a *tensor of Euler curvature*.

Consider now a field of normal vectors Y^α. Its absolute differential has the tangent component

$$Y^\lambda \omega_\lambda^i$$

and the normal component

$$dY^\alpha + Y^\lambda \omega_\lambda^\alpha.$$

For normal vectors, the forms ω_λ^α play the same role as the forms ω_i^j for tangent vectors. The normal component is called the *interior normal differential*.

Thus, there are two (invariant) differentiations, and in addition, the Euler curvature gives the component (normal or tangent) of the absolute variation of a vector (normal or tangent).

155. Tensor character of Euler's curvature. Consider the scalar product of the vectors **Y** and D**X**:

$$\mathbf{Y} \cdot D\mathbf{X} = Y_\lambda X^k \omega_k^\lambda.$$

This product explains the tensorial character of the linear forms ω_k^λ. But here the tensor has "two faces" (two-sided): one index is related to the tangent displacement, and the second one to the normal displacement.

Thus, we come to the necessity of considering the tensors of the form

$$a_{ijk\ldots|\alpha\beta\gamma\ldots}.$$

In the case $n = 3$, we had only one normal index which we lowered. The quantities $\gamma_{ij}{}^\alpha$ form a tensor of this kind. For such tensors, we can define interior differentiation by setting, for example,

$$d(a_{i\alpha} X^i Y^\alpha) = Da_{i\alpha} \cdot X^i Y^\alpha.$$

If a tangent vector X^i and a normal vector Y^α are transported by parallelism, then

$$dX^i = -X^k \omega_k^i,$$

and

$$dY^\alpha = -Y^\lambda \, \omega_\lambda^\alpha.$$

Therefore, we have

$$d(a_{i\alpha} X^i Y^\alpha) = da_{i\alpha} X^i Y^\alpha + a_{i\alpha} \, dX^i \, Y^\alpha + a_{i\alpha} X^i \, dY^\alpha$$

$$= da_{i\alpha} X^i Y^\alpha - a_{i\alpha} Y^\alpha X^k \, \omega_k^i - a_{i\alpha} X^i Y^\lambda \, \omega_\lambda^\alpha$$

$$= da_{i\alpha} X^i Y^\alpha - a_{k\alpha} Y^\alpha X^i \, \omega_i^k - a_{i\lambda} X^i Y^\alpha \, \omega_\alpha^\lambda$$

$$= (da_{i\alpha} - a_{k\alpha} \, \omega_i^k - a_{i\lambda} \, \omega_\alpha^\lambda) X^i Y^\alpha = Da_{i\alpha} X^i Y^\alpha.$$

Since the vectors X^i and Y^α are arbitrary, it follows that

$$Da_{i\alpha} = da_{i\alpha} - a_{k\alpha} \, \omega_i^k - a_{i\lambda} \, \omega_\alpha^\lambda. \tag{25.4}$$

Similar to interior differentiation, exterior differentiation of a linear tensor form $\tilde{\omega}_i^\alpha$ is defined by means of the formula

$$D\tilde{\omega}_i^\alpha = d\tilde{\omega}_i^\alpha + \omega_i^k \wedge \tilde{\omega}_k^\alpha + \omega_\lambda^\alpha \wedge \tilde{\omega}_i^\lambda. \tag{25.5}$$

156. The second system of structure equations. Consider the second system of structure equations:

$$d\omega_i^j = \omega_i^k \wedge \omega_k^j + \omega_i^\lambda \wedge \omega_\lambda^j + R_{ij,kh} \, \omega^k \wedge \omega^h, \tag{25.6}$$

$$d\omega_\alpha^\beta = \omega_\alpha^\lambda \wedge \omega_\lambda^\beta + \omega_\alpha^k \wedge \omega_k^\beta + R_{\alpha\beta,kh} \, \omega^k \wedge \omega^h, \tag{25.7}$$

$$d\omega_i^\alpha = \omega_i^k \wedge \omega_k^\alpha + \omega_i^\lambda \wedge \omega_\lambda^\alpha + R_{i\alpha,kh} \, \omega^k \wedge \omega^h. \tag{25.8}$$

The internal Riemannian curvature of the submanifold V (up to sign) is

$$D\omega_i^j = d\omega_i^j - \omega_i^k \wedge \omega_k^j.$$

By the first structure equation (25.6), this curvature is equal to the sum of

1) the Riemannian curvature of the ambient manifold in the same planar direction, and

2) the expression

$$\omega_i^\lambda \wedge \omega_\lambda^j,$$

which depends only on the Euler curvature of the submanifold V.

By analogy with the case $n = 3$, we shall call this curvature the *Gaussian curvature* (up to sign).

The components of the Gaussian curvature are determined by the expression

$$\omega_i^\lambda \wedge \omega_j^\lambda.$$

Equation (25.8) shows that the invariant exterior differential of the Euler curvature ω_i^α depends only on the Riemannian curvature of the ambient manifold.

Since

$$\omega_i^\alpha = \gamma_{ij}{}^\alpha \, \omega^j,$$

this exterior differential

$$d\omega_i^\alpha = \gamma_i{}^\alpha{}_{[jk]} \, \omega^j \wedge \omega^k, \quad \gamma_i{}^\alpha{}_{[jk]} = \gamma_i{}^\alpha{}_{jk} - \gamma_i{}^\alpha{}_{kj},$$

depends only on the skew-symmetric part of the derivative tensor of the Euler curvature.

As for equation (25.7), it is interpreted similarly to equation (25.6): the bivector

$$d\omega_\alpha^\beta - \omega_\alpha^\lambda \wedge \omega_\lambda^\beta,$$

called (up to sign) the *Riemannian torsion* of the submanifold, is a rotation which the normal vector undergoes after a parallel transport along a cycle on the manifold. The term

$$\omega_\alpha^k \wedge \omega_k^\beta,$$

depending only on the Euler curvature, defines the *Gaussian torsion*.

PARTICULAR CASES

A HYPERSURFACE IN A FOUR-DIMENSIONAL SPACE

157. Principal directions and principal curvatures. There is only one normal index 4. The Euler curvature is defined by the forms ω_i^4. We can set

$$\omega_i^4 = a_{ik} \omega^k, \quad a_{ik} = a_{ki}. \tag{25.9}$$

The symmetric tensor a_{ik} allows us to define the quadratic form

$$\varphi = \omega^i \omega_i^4 = a_{ij} \omega^i \omega^j. \tag{25.10}$$

The expression

$$\frac{\varphi}{ds^2}$$

is the *normal curvature* of a line on the hypersurface V.

The three axes of the frame, for which

$$\varphi = \frac{(\omega^1)^2}{R_1} + \frac{(\omega^2)^2}{R_2} + \frac{(\omega^3)^2}{R_3},$$

are called the *principal directions*. The coefficients $\dfrac{1}{R_i}$ are called the *principal curvatures*. The *Gaussian curvature* is defined by the forms

$$\omega_i^4 \wedge \omega_j^4 = \frac{1}{R_i R_j}\, \omega^i \wedge \omega^j \quad (i, j \text{ not summed}).$$

Thus, we have three *principal Gaussian curvatures*

$$\frac{1}{R_2 R_3}, \quad \frac{1}{R_3 R_1}, \quad \frac{1}{R_1 R_2}.$$

158. Hypersurface in the Euclidean space. Consider a hypersurface in the four-dimensional *Euclidean* space. The principal Gaussian curvatures become the Riemannian curvatures. If one deforms the hypersurfaces in such a way that its line element ds^2 is preserved, then

$$\frac{1}{R_2 R_3} = \frac{1}{R_2' R_3'}, \quad \frac{1}{R_3 R_1} = \frac{1}{R_3' R_1'}, \quad \frac{1}{R_1 R_2} = \frac{1}{R_1' R_2'},$$

i.e.,

$$\frac{1}{R_1} = \frac{1}{R_1'}, \quad \frac{1}{R_2} = \frac{1}{R_2'}, \quad \frac{1}{R_3} = \frac{1}{R_3'},$$

(possibly except signs—different signs would correspond to symmetry with respect to a certain hyperplane).

Thus, deformable hypersurfaces are equal or symmetric. In general, it is impossible to deform a hypersurface without changing ds^2. This is the *theorem of Beez*.[1]

Given a quadratic form of three variables, is there a hypersurface in a four-dimensional Euclidean space corresponding to this form? A three-dimensional Riemannian submanifold (a hypersurface) depends on three functions of three variables (coefficients in a decomposition of the line element into a sum of squares). But, in a four-dimensional space, a hypersurface is defined just by one function of three variables

$$x_4 = f(x_1, x_2, x_3).$$

Thus, in general, the problem does not have a solution.

In order to solve the problem, we need to take a space of six dimensions. If

$$ds^2 = g_{ij}\, du^i\, du^j, \quad i, j = 1, 2, 3,$$

then comparing the coefficients for all combinations (with repetitions) of $du^i\, du^j$ in the equation

$$\sum_{k=1}^{6} (dx_k)^2 = g_{ij}\, du^i\, du^j,$$

---

[1]Beez, R. *Zur Theorie des Krümmungsmasses von Mannigfaltigkeiten höherer Ordnung*, Schlmlich Z., **XX** (1875), 423–444 (JFM 7, p. 306); cf. Souvarow, Bull. Science. Math., 1873. (?)

where x_k are considered as functions of u^1, u^2, u^3, we obtain six equations for six unknown functions x_k with respect to three independent variables u^i. In general, these equations admit a solution.

In general, the line element ds^2 with differentials of p parameters belongs to a submanifold of p dimensions in the Euclidean space of dimension $\frac{p(p+1)}{2}$.

TWO-DIMENSIONAL SURFACES
IN A FOUR-DIMENSIONAL MANIFOLD

159. Ellipse of normal curvature. There are two tangent indices, 1 and 2, and two normal indices, 3 and 4. The Euler curvature is defined by the forms

$$\omega_1^3, \ \omega_1^4, \ \omega_2^3, \ \omega_2^4,$$

or by two quadratic forms,

$$\varphi^3 = \omega^1 \, \omega_1^3 + \omega^2 \, \omega_2^3 = a_{11} \, (\omega^1)^2 + 2a_{12} \, \omega^1 \, \omega^2 + a_{22} \, (\omega^2)^2 \qquad (25.11)$$

and

$$\varphi^4 = \omega^1 \, \omega_1^4 + \omega^2 \, \omega_2^4 = b_{11} \, (\omega^1)^2 + 2b_{12} \, \omega^1 \, \omega^2 + b_{22} \, (\omega^2)^2. \qquad (25.12)$$

Here we find the *vector of normal curvature*

$$\overrightarrow{\left(\frac{1}{\rho_n} \right)} = \frac{\varphi^3 \, \mathbf{I}_3 + \varphi^4 \, \mathbf{I}_4}{ds^2}; \qquad (25.13)$$

in this formula $\dfrac{\varphi^3}{ds^2}$ and $\dfrac{\varphi^4}{ds^2}$ are the components of the normal curvature relative to the two axes normal to the curve.

Denote by θ the angle formed by the tangent to the curve with the first axis in the tangent plane $(\mathbf{I}_1, \mathbf{I}_2)$. Then we have

$$\omega^1 = \cos\theta \, ds, \ \ \omega^2 = \sin\theta \, ds. \qquad (25.14)$$

Suppose that $\overrightarrow{MP}$ is the vector of normal curvature, and x and y are the coordinates of a point P in the normal plane $(\mathbf{I}_3, \mathbf{I}_4)$ (Fig. 22). Then

$$x = \frac{\varphi^3}{ds^2}, \ \ y = \frac{\varphi^4}{ds^2},$$

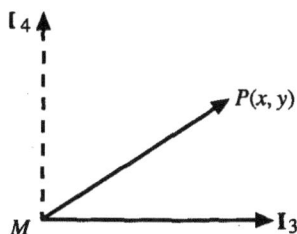

Fig. 22

i.e.,

$$
\begin{cases}
x = a_{11} \cos^2 \theta + 2a_{12} \cos \theta \sin \theta + a_{22} \sin^2 \theta, \\
y = b_{11} \cos^2 \theta + 2b_{12} \cos \theta \sin \theta + b_{22} \sin^2 \theta,
\end{cases}
$$

or

$$
\begin{cases}
x = \dfrac{a_{11} + a_{22}}{2} + \dfrac{a_{11} - a_{22}}{2} \cos 2\theta + a_{12} \sin 2\theta, \\
y = \dfrac{b_{11} + b_{22}}{2} + \dfrac{b_{11} - b_{22}}{2} \cos 2\theta + b_{12} \sin 2\theta.
\end{cases}
\tag{25.15}
$$

Thus, the locus of points $P(x, y)$ is an ellipse centered at $H\left(\frac{a_{11}+a_{22}}{2}, \frac{b_{11}+b_{22}}{2}\right)$. This ellipse is called the *ellipse of indicatrix of normal curvature.*

The vector $\overrightarrow{MH}$ is called the *vector of mean curvature* (Bompiani),

$$
\overrightarrow{MH} = \frac{a_{11} + a_{22}}{2} \mathbf{I}_3 + \frac{b_{11} + b_{22}}{2} \mathbf{I}_4.
\tag{25.16}
$$

This vector is defined by the contracted tensor of the Euler curvature.

In order to find the Euler curvature, four essential constants must be given. Take the vector $\mathbf{I}_1$ to be parallel to the major axis of the ellipse of indicatrix centered at $H(\alpha, \beta)$ and having the semiaxes a and b. Suppose that $\theta = 0$ at the endpoint of the major axis of the ellipse. Then the ellipse has the parametric equations

$$
\begin{cases}
x = \alpha + a \cos 2\theta, \\
y = \beta + b \sin 2\theta.
\end{cases}
\tag{25.17}
$$

Thus, we have

$$
\begin{cases}
\dfrac{a_{11} + a_{22}}{2} = \alpha, \quad \dfrac{b_{11} + b_{22}}{2} = \beta, \\[2mm]
\dfrac{a_{11} - a_{22}}{2} = a, \quad a_{12} = 0, \\[2mm]
\dfrac{b_{11} - b_{22}}{2} = 0, \quad b_{12} = b,
\end{cases}
\tag{25.18}
$$

and the quadratic forms φ^3 and φ^4 are

$$
\begin{cases}
\varphi^3 = (\alpha + a)\,(\omega^1)^2 + (\alpha - a)\,(\omega^2)^2, \\[2mm]
\varphi^4 = \beta\,(\omega^1)^2 + 2b\,\omega^1\,\omega^2 + \beta\,(\omega^2)^2.
\end{cases}
\tag{25.19}
$$

It follows that

$$
\begin{cases}
\omega_1^3 = (\alpha + a)\,\omega^1, \quad \omega_1^4 = \beta\,\omega^1 + b\,\omega^2, \\[2mm]
\omega_2^3 = (\alpha - a)\,\omega^2, \quad \omega_2^4 = b\,\omega^1 + \beta\,\omega^2,
\end{cases}
\tag{25.20}
$$

and

$$
\omega_1^3 \wedge \omega_2^3 + \omega_1^4 \wedge \omega_2^4 = (\alpha^2 - a^2)\,\omega^1 \wedge \omega^2 + (\beta^2 - b^2)\,\omega^1 \wedge \omega^2.
$$

Therefore, the Gaussian curvature is

$$
\alpha^2 + \beta^2 = a^2 + b^2.
\tag{25.21}
$$

This is the power of the point M with respect to the orthoptic circle of the ellipse. In fact, the *orthoptic circle* is the locus of points from which an ellipse is seen at a straight angle. The tangents to the ellipse at its vertices definitely intersect each other at a straight angle and define a rectangle whose vertices lie on the orthoptic circle. The center of the orthoptic circle coincides with the center (α, β) of the ellipse, and its radius is equal to half the diagonal of the rectangle, i.e., it is equal to $\sqrt{a^2 + b^2}$. The power of the point M with respect to the circle is equal to the product of the entire secant from the point M and its exterior part. The entire secant is equal to the sum of the distance from the point M to the center of the circle and the radius, and its exterior part is equal to the difference of these two segments.

Thus, the desired power is equal to the product

$$
\left(\sqrt{\alpha^2 + \beta^2} + \sqrt{a^2 + b^2} \right)\left(\sqrt{\alpha^2 + \beta^2} - \sqrt{a^2 + b^2} \right) = \alpha^2 + \beta^2 - (a^2 + b^2).
$$

In the three-dimensional space, the indicatrix reduces to the segment $P_1 P_2$ consisting of two parts, $MP_1 = \frac{1}{R_1}$ and $MP_2 = \frac{1}{R_2}$, and the Gaussian curvature, or the total curvature, is the power of the point M with respect to the orthoptic circle (the segment $P_1 P_2$).

We turn now to the Gaussian curvature. We have

$$\omega_1^3 \wedge \omega_1^4 + \omega_2^3 \wedge \omega_2^4 = 2a\,b\,\omega^1 \wedge \omega^2. \tag{25.22}$$

Hence, the Gaussian curvature is $2ab$. Note that if we assume that $a > 0$, we must take $b > 0$, if the ellipse is described in the positive direction, when θ increases from 0 to π.

160. Generalization of classical notions. CURVATURE LINES. We can consider the curvature lines as the lines that correspond to the endpoints of the major axis of the ellipse. However, more often, the curvature line is defined as a line for which the tangential component of the absolute displacement of a normal vector is a tangent to this line.

Thus, for such a line, we have

$$\begin{cases} \omega^1\,\omega_2^3 - \omega^2\,\omega_1^3 = 0, \\ \omega^1\,\omega_2^4 - \omega^2\,\omega_1^4 = 0. \end{cases}$$

We find a condition for existence of the curvature lines. Substituting the values of the forms ω_i^α from equations (25.20) into the above equations, we find that

$$\begin{cases} -2a\,\omega^1\,\omega^2 = 0, \\ b\,[(\omega^1)^2 - (\omega^2)^2] = 0. \end{cases}$$

These equations are inconsistent if both a and b are zero. The condition for existence of the curvature lines is the vanishing of either a or b, i.e., the vanishing of the Gaussian curvature (25.22). In this case, there are two families of orthogonal curvature lines.

ASYMPTOTIC LINES. An asymptotic line is a curve with zero normal curvature. For its existence on the surface, it is necessary that the ellipse of indicatrix of the normal curvature passes through the point M. Then there exists one family of asymptotic lines. There will be two families of asymptotic lines if the ellipse of indicatrix degenerates into a double straight line.

If asymptotic lines exist, then the Gaussian curvature is negative (since the point M is an interior point of the orthoptic circle).

Consider the case when the ellipse of indicatrix is a double straight line situated on the axis $\mathbf{I}_3$. Then the form φ_4 is identically zero. The surface is a hyperplane, i.e., it is a surface of a three-dimensional Euclidean space.

In fact, exterior differentiation of the equations

$$\omega_1^4 = 0, \quad \omega_2^4 = 0$$

(they follow from the condition $\varphi_4 = 0$) gives

$$\omega_1^3 \wedge \omega_3^4 = 0, \quad \omega_2^3 \wedge \omega_3^4 = 0.$$

It follows that the linear form ω_3^4 is simultaneously a multiple of two linearly independent forms ω_1^3 and ω_2^3. Hence, this form vanishes, $\omega_3^4 = 0$. As a result, we have

$$d\mathbf{I}_4 = 0,$$

the direction $\mathbf{I}_4$ is fixed, and the point M describes a hyperplane and contains our surface.

161. Minimal surfaces. Suppose that the mean curvature (25.16) is zero. By (25.18), this implies that

$$\alpha = 0, \quad \beta = 0.$$

Hence, the point H coincides with the point M. Then it follows from (25.19) that

$$\varphi^3 = a \left[(\omega^1)^2 - (\omega^2)^2 \right],$$

$$\varphi^4 = 2b\omega^1 \omega^2.$$

The Gaussian curvature (25.21), which is now equal to

$$-a^2 - b^2,$$

is negative.

This property characterizes a minimal surface, i.e., a surface of minimal area for a given contour.

In fact, suppose that the symbols d and δ correspond respectively to a displacement along the surface and a displacement from one surface to another, infinitesimally close, surface.

If we assume that the latter displacement is orthogonal to the surface, then obviously the forms $\omega_i^k(\delta) = e_i^k$ satisfy the equations $e^1 = 0$, $e^2 = 0$ on the surface and $e^3 = 0$, $e^4 = 0$ on the contour.

Now the variation of the surface area is

$$\delta \int \omega^1 \wedge \omega^2 = \int \delta\omega^1 \wedge \omega^2 - \int \omega^1 \wedge \delta\omega^2.$$

But from the structure equation

$$d\omega^1 = \omega^2 \wedge \omega_2^1 + \omega^\alpha \wedge \omega_\alpha^1,$$

we obtain

$$\delta\omega^1 = -e_2^1 \omega^2 + e^\alpha \omega_\alpha^1,$$

and similarly

$$\delta\omega^2 = -e_1^2 \omega^1 + e^\alpha \omega_\alpha^2.$$

Therefore,

$$\delta \int \omega^1 \wedge \omega^2 = \int e^\alpha (\omega^2 \wedge \omega_1^\alpha - \omega^1 \wedge \omega_2^\alpha), \quad \alpha = 3, 4.$$

Thus, a necessary condition for the minimum of the surface area (the vanishing of the first variation) is

$$\omega^1 \wedge \omega_2^\alpha - \omega^2 \wedge \omega_1^\alpha = 0.$$

If we set

$$\begin{cases} \omega_1^\alpha = a_{11}^\alpha \omega_1 + a_{12}^\alpha \omega_2, \\ \omega_2^\alpha = a_{12}^\alpha \omega_1 + a_{22}^\alpha \omega_2, \end{cases}$$

we obtain

$$a_{11}^\alpha + a_{22}^\alpha = 0, \quad \alpha = 3, 4,$$

or using notation (25.18), we have

$$\alpha = 0, \quad \beta = 0.$$

If we deform the surface by a normal displacement orthogonal to the straight line MH, then the variation of the surface area is zero.

162. Finding minimal surfaces. We shall say that a function V is *harmonic* if the second differential parameter of this function vanishes,

$$\Delta_2 V = 0.$$

In n-dimensional Euclidean space, the Cartesian coordinates of a point of a minimal surface are harmonic functions on the surface.

In fact, the point coordinates can be represented by an analytic point $\mathbf{M}$. The coefficients of the forms ω^i in the decomposition

$$dV = V_1 \omega^1 + V_2 \omega^2 + \ldots + V_n \omega^n$$

are said to be the components of the gradient. Thus, for

$$d\mathbf{M} = \omega^1 \mathbf{I}_1 + \omega^2 \mathbf{I}_2,$$

the components of the gradient are the vectors $\mathbf{I}_1$ and $\mathbf{I}_2$. In a space of constant curvature, we obtain the formulas

$$D\mathbf{M}_1 = D\mathbf{I}_1 = d\mathbf{I}_1 + \mathbf{I}_2 \omega_1^2 = \omega_1^\alpha \mathbf{I}_\alpha - K \omega_1 \mathbf{M} = \mathbf{M}_{11} \omega_1 + \mathbf{M}_{12} \omega_2,$$

$$D\mathbf{M}_2 = D\mathbf{I}_2 = d\mathbf{I}_2 + \mathbf{I}_1 \omega_2^1 = \omega_2^\alpha \mathbf{I}_\alpha - K \omega_2 \mathbf{M} = \mathbf{M}_{21} \omega_1 + \mathbf{M}_{22} \omega_2,$$

similar to formulas (24.32). Comparing the last two columns, we find that

$$\mathbf{M}_{11} = a_{11}^\alpha \mathbf{I}_\alpha - K \mathbf{M},$$

$$\mathbf{M}_{22} = a_{22}^\alpha \mathbf{I}_\alpha - K \mathbf{M},$$

and
$$\Delta_2 \, M = M_{11} + M_{22} = (a_{11}^{\alpha} + a_{22}^{\alpha}) \, I_{\alpha} - 2K \, M.$$

If the surface is minimal, then
$$\Delta_2 \, M + 2K \, M = 0.$$

If $K = 0$, then we find the equation
$$\Delta_2 \, M = 0,$$

characterizing the minimal surfaces.

Suppose that the form
$$ds^2 = E \, du^2 + 2F \, du \, dv + G \, dv^2$$

reduces to the form
$$A^2 \, (du^2 + dv^2).$$

Then
$$\omega^1 = A \, du, \quad \omega^2 = A \, dv.$$

The components of $\overrightarrow{\mathrm{grad}\, V}$ are
$$V_1 = \frac{1}{A} \frac{\partial V}{\partial u}, \quad V_2 = \frac{1}{A} \frac{\partial V}{\partial v},$$

and
$$V_1 \, \omega^2 - V_2 \, \omega^1 = \frac{\partial V}{\partial u} \, dv - \frac{\partial V}{\partial v} \, du$$

is the element of the flux of the gradient. This flux is an exact differential if V is a harmonic function, since
$$\int_C (V_1 \, \omega^2 - V_2 \, \omega^1) = \iint_S \Delta_2 V \, \omega^1 \wedge \omega^2.$$

Thus,
$$\frac{\partial^2 V}{\partial u^2} + \frac{\partial^2 V}{\partial v^2} = 0,$$

and the function V is harmonic with respect to the coordinates u and v.

If V is a function of a complex variable, which is complex conjugate to the function f, then we can define a minimal surface by the equations
$$x_i = f_i(u + i\,v) + \varphi_i(u - i\,v).$$

The functions f_i must be chosen in such a way that
$$ds^2 = A^2 \, (du^2 + dv^2).$$

We have
$$ds^2 = A^2 \, (du + i\,dv)(du - i\,dv) = A^2 \, (du^2 + dv^2)$$

and

$$dx_i = f_i'(du + i\,dv) + \varphi_i'(du - i\,dv).$$

Therefore, it is necessary to have

$$\sum_{i=1}^{3}(f_i')^2 = 0.$$

This is an extension of a classical result to the case of four dimensions.

If we substitute if_k for f_k, we obtain the minimal surface *associated* with the given one. The theory of associated surfaces can be extended to the case of n dimensions. The tangent planes at corresponding points are parallel, and the corresponding tangents are perpendicular.

Subject Index

www.ingramcontent.com/pod-product-compliance
Lightning Source LLC
Chambersburg PA
CBHW060238220326
41598CB00027B/3976